Laboratory Manual for
LIBERAL ARTS
PHYSICS

Second Edition

Art Hobson
University of Arkansas

Marie Baehr
Elmhurst College

Earl C. Swallow
Elmhurst College

Prentice
Hall

Pearson Education, Inc.
Upper Saddle River, NJ 07458

Assistant Editor: Christian Botting
Senior Editor: Erik Fahlgren
Editor-In-Chief: John Challice
Executive Managing Editor: Kathleen Schiaparelli
Assistant Managing Editor: Dinah Thong
Production Editors: Dana Dunn and Christy Wrightington
Art Director: Jayne Conte
Cover Designer: Karen Salzbach
Manufacturing Buyer: Ilene Kahn
Cover Photograph: Light Bulb, Getty Images/PhotoDisc

© 2003 Pearson Education, Inc.
Pearson Education, Inc.
Upper Saddle River, NJ 07458

Printed in the United States of America

10 9 8 7 6 5 4 3 2 1

ISBN 0-13-101107-3

Pearson Education Ltd., *London*
Pearson Education Australia Pty. Ltd., *Sydney*
Pearson Education Singapore, Pte. Ltd.
Pearson Education North Asia Ltd., *Hong Kong*
Pearson Education Canada, Inc., *Toronto*
Pearson Educacíon de Mexico, S.A. de C.V.
Pearson Education—Japan, *Tokyo*
Pearson Education Malaysia, Pte. Ltd.
Pearson Education, Inc., *Upper Saddle River, New Jersey*

For Claire Bartels

CONTENTS

Experiment

PREFACE

This laboratory manual is specifically designed for use in a physics course for non-science students. It is particularly suitable for use with the latest edition of the textbook *Physics: Concepts and Connections* by Art Hobson published by Prentice Hall in 2002. It is also appropriate for use with other *liberal arts physics* texts or *conceptual physics* texts. We have sought to provide a substantial set of laboratory exercises specifically suited in both content and structure to a modern course for non-science students.

In addition to the usual emphases on illustrating physical principles and providing experience with physical phenomena, these lab exercises have been constructed to show some of the ways in which systematic experimental observation can lead to insight and understanding. We have avoided the ingenuous pretense that important physical principles can be "tested" in an elementary course laboratory. In terms of buzzwords that enjoy some currency, we have intentionally attended to *process* (sometimes explicitly, sometimes implicitly) in addition to *content* in these labs. Giving non-science students a clearer picture of "how we know" is a crucial function of any liberal arts physics course.

We have included a large enough number of exercises that professors can select labs to match the emphases of their particular course. Some of the labs focus on physical principles, concepts, or phenomena; others, on issues of energy and environment; still others, on understanding scientific data, experimental errors, and "numeracy." Although most of the labs deal with topics explicitly discussed in *Physics: Concepts and Connections*, some provide extensions or elaborations. The labs can be assigned in a variety of orders to fit differing course structures, with the exception of the first few labs that help to develop necessary laboratory skills.

Because the physics laboratory provides an excellent opportunity for students to work together, nearly all the exercises are structured so that they can be performed in a group. Several of them expressly require group work, sometimes requiring the pooling of data from the entire laboratory class.

Although this manual can be used effectively without data acquisition computers in the laboratory, many of the exercises in this edition have been structured to take advantage of computers where they are available. In this edition, the possibility of using spreadsheets or graphing calculators has been included wherever it seemed appropriate. Several exercises (#10, #18, #22, #23, and #28) explicitly call for computerized sensors. Several others (#1, #5-9, #11, #30, #31) include provision for using a spreadsheet or graphing calculator.

One laboratory in the previous edition that dealt with air pollution has been eliminated because the use of smoke was in conflict with institutional regulations at almost all U.S. colleges and universities. In this edition, many of the exercises have been modified to provide greater clarity and enhanced intellectual engagement.

The list of **BASE CONCEPTS** at the beginning of each lab is intended to indicate relevant ideas with which students should have some familiarity. It is *not necessary* for the primary topic of every lab to be discussed in a classroom lecture before the lab is carried out. It is perfectly reasonable and desirable for students to meet some topics for the first time in the laboratory, though doing this will usually require considerable extra effort and care on the part of the lab instructor.

If any of the **BASE CONCEPTS** seems unfamiliar to you as a student, you will probably find it useful (and educational) to look it up in a textbook or a reference book. We have quite intentionally **not** attempted to make this lab manual an encyclopedic document. We hope this will encourage you to refer to other sources — just as a serious investigator in any area of study must.

Questions play a central role in the practice of experimental physics. We have attempted to capture this spirit of questioning by including a variety of questions (sequenced alphabetically with uppercase letters) throughout each lab exercise. In some cases there are **INITIAL QUESTIONS** that set the stage for the investigation before any discussion. Often there are interpretive and predictive questions embedded in the **INTRODUCTION** and **PROCEDURE** sections. We point these out by marking them with a ➤ graphic symbol. Finally, there are **GLOBAL QUESTIONS** at the end of many exercises. These involve broader reflection on the content and results.

All these questions should be answered as they are encountered. They are intended to give you, the student, a sense of the variety of questions a practicing scientist might pose to himself or herself. In actual experimental research, the questions are far more numerous. Some are posed, refined, and addressed in formal written documents (proposals, design reports, memos, etc.). Others are simply part of the experimenter's personal thought processes and notes. All of them play a crucial role in converting observations and facts into insight and knowledge. In this same spirit we urge you not to be satisfied with our questions, pose and answer additional questions of your own as you carry out your experiments. Also in the spirit of experimental physics, have fun and learn — that's one of the best parts of life!

Marie Baehr
Earl C. Swallow

ACKNOWLEDGMENTS

We are indebted to many colleagues, students, and friends for their helpful comments and encouragement . Special thanks go to Kirk Lentz, now a chemistry teacher at Prospect High School, who carefully carried out each experiment and offered myriad helpful suggestions and comments, and to our Physics 101, 121, and 122 students at Elmhurst College (EC), particularly Rita Mock Gillespie and Anna Baughman. Stimulating discussions with Physics 101 laboratory assistants Mark Hall, Molly Giblin, Ray Flynn, and Dan Carnduff are gratefully acknowledged. Workshops and writings by Priscilla Laws and Ron Thornton contributed greatly to the style of this manual and provided ideas that proved adaptable to the setting of the liberal arts physics course. Teaching Institutes led by Dan Apple of Pacific Crest made important contributions to our understanding and appreciation of the value of strengthening each student's learning *processes* while seriously attending to *content*. Discussions concerning the EC general education program encouraged us to maintain our interest in this project. Continued prodding from Christian Botting, our new Prentice Hall editor, has kept us moving forward. Kind assistance "above and beyond" was provided by Rita Andreuccetti, secretary to the EC Physics Department, in preparing this new edition. Finally, we thank our families and life companions who have given us support and understanding through many hours of distraction and effort.

In the first edition we said, "Typing, reading, checking, correcting, producing and modifying drawings, making suggestions, trying type faces, providing encouragement, and fending off departmental distractions: all of these were done by Claire Bartels, secretary to the Physics, Mathematics, and Computer Science Departments at Elmhurst. We appreciate her continuing help and friendship more than we can say." Unfortunately, due to her untimely death, we must now say that we miss her more than we can say. We respectfully dedicate this edition to her.

We invite any and all comments, corrections, additions, and suggestions for further improving this laboratory manual.

Marie Baehr
email: marieb@elmhurst.edu

Earl C. Swallow
email: earls@elmhurst.edu

Department of Physics
Elmhurst College
190 Prospect Avenue
Elmhurst, Illinois 60126

It does not very much matter to science what these vague speculations lead to, for meanwhile it forges ahead in a hundred directions, in its own precise experimental way of observation, widening the bounds of the charted region of knowledge, and changing human life in the process. It may be on the verge of discovering vital mysteries, and yet they may elude it. Still it will go on along its appointed path, for there is no end to its journeying. Ignoring for the moment the Why? of philosophy, it will go on asking How?, and as it finds this out it gives greater content and meaning to life, and perhaps takes us some way to answering the Why.

> – Jawaharlal Nehru
> First prime minister of India

EXPERIMENT 1

TWO-DIMENSIONAL GRAPHING

Name_____   Partner(s)_____

Date_____   _____

Section Number_____   _____

BASE CONCEPTS: Graph Reading

INTRODUCTION:

This lab deals with graphing in two dimensions. In all cases the graph will consist of one horizontal axis and one vertical axis. The horizontal axis is used for the independent variable (or quantity) and typically is called the x-axis or domain. The vertical axis is used for the dependent variable (or quantity) and typically is called the y-axis or ordinate. To locate any point on a graph, two numbers need to be known: the value of the horizontal axis variable and the *corresponding* value of the vertical axis variable. From a mathematical point of view, these two numbers constitute an ordered pair.

If the y-axis variable of the ordered pair can be generated from the x-axis variable of the pair by an equation in the form $y = mx + b$, where m and b are both constants (i.e., their values do not change as the values of x and y change), then we say that the y-axis variable is a *linear function* of the x-axis variable, because when values are plotted, a straight line can be drawn though all the ordered pairs. In this lab you will investigate the graphs of different linearly related quantities.

PROCEDURE AND QUESTIONS:

1. First, look at Figure 1, a graph of Celsius temperature (the vertical axis variable) as a function of Fahrenheit temperature (the horizontal axis variable). Examination of the graph indicates that these two quantities do appear to be linearly related, so we need to find the slope (m) and the intercept (b) in the equation

$$t[°C] = m\, t[°F] + b,$$ Eq. 1

where $t[°C]$ is the Celsius temperature, and $t[°F]$ is the Fahrenheit temperature.

A. Examining the graph in Figure 1, determine the Celsius temperature when the Fahrenheit temperature is 0°F.

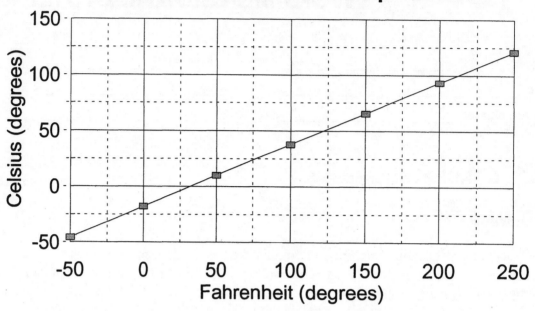

Celsius vs. Fahrenheit Temperatures

Figure 1

Table 1
Data (Ordered Pairs) for Figure 1 Graph

°F	− 50.00	0.00	50.00	100.00	150.00	200.00	250.00
°C	− 45.56	− 17.78	10.00	37.78	65.56	93.33	121.11

B.   Using the ordered pair [0, y], where y is the value found in Question A, write Equation 1, using 0, y, m, and b.

C.   From your answers for Question A, what is the numerical value of b for this set of data?

D.   From the graph, what is the Celsius temperature when the Fahrenheit temperature is 200°F?

E.   Using Equation 1 and answers from Questions C and D, determine the value of the slope m,

where $m = \dfrac{t[°C] - b}{200°F - 0°F}$. In this equation, $t[°C]$ is the ordered pair "mate" of 200°

Fahrenheit, found as the answer to Question D.

From answering these questions you should have gotten a glimpse of the meaning of b: it is the value of the vertical axis variable when the horizontal axis variable is 0. For this reason, b is called the *y-intercept*. Although you have found a value for the slope (the conventional name for m), it may not have much physical meaning to you yet. To develop a greater understanding of the physical interpretation of the slope, look at the graph and answer the following questions.

F.   What is the value of $t[°C]$ when $t[°F] = 200°F$?

G.   What is the value of $t[°C]$ when $t[°F] = 150°F$?

H.   What is the value of $t[°C]$ when $t[°F] = 100°F$?

I.   What is the value of $t[°C]$ when $t[°F] = 50°F$?

J.   What is the value of $t[°C]$ when $t[°F] = 0°F$?

K.   Using any two of the answers from F through J, take the difference of the two Celsius temperatures and divide by the difference of the two corresponding Fahrenheit temperatures. For example, if you were to do the calculations for the values in G and H, your answer would

be $\dfrac{t[°C] \text{ in G} - t[°C] \text{ in H}}{150°F - 100°F}$. Make sure that you keep the proper elements of the ordered

pairs in the numerator and the denominator.

L. Repeat Question K three times, using different sets of numbers from F through J each time.

M. Compare the values calculated in K with L to the value of the slope calculated in E. Considering the graph (Figure 1), why should these three values be the same?

N. In your own words, for a given vertical axis variable (y) and horizontal axis variable (x) that are linearly related, state how you would find the slope.

Once you have the equation that describes the data, you can use it to predict values. For example, suppose you would like to know the Celsius temperature equivalent to 98.6°F.

O.   Using the values for the intercept b and the slope m you obtained, calculate $t[°C]$ from $t[°C] = m \times 98.6°F + b$, replacing m and b with your calculated values.

P.   Using the graph, estimate $t[°C]$ for $t[°F] = 98.6°F$. Compare this value with that found in Question O.

2.   *Experimental* data points often have large enough errors that even if the two quantities are linearly related, the data points do not fall *exactly* on a straight line. When this is the case, you can draw a straight line so that some of the data points lie above the line and some lie below the line. You must be careful though. If the points that lie below the line are all adjacent to one another, as are the points above the line, you probably have graphed variables that are not linearly dependent. If, however, the points below the line and the points above the line are randomly distributed, then the variables are probably linearly dependent. See Figures 2 and 3 for examples.

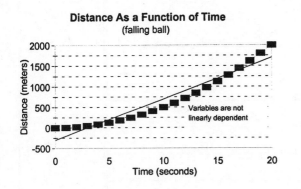

Figure 2

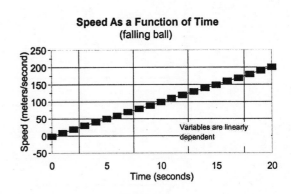

Figure 3

Table 2 contains some data that you are to graph by hand or on a graphics calculator. After graphing the data points, draw a straight line that best fits the data points. One of the easiest ways to do this is to make sure you have as many data points above the line as you do below the line and that the points above and below are approximately randomly distributed along the line (similar to Figure 3). If you use a graphics calculator, draw a regression line and display the slope and intercept of the regression line. (The calculator creates the regression line by calculating the best fit straight line)

Q.   After drawing the best-fit straight line, find the slope and intercept of the line and write an equation to relate the vertical axis variable and the horizontal axis variable.

Table 2
Speed of Dropped Ball As a Function of Time

TIME OF FALL (seconds)	SPEED OF BALL (meters/second)
0	0
1	10
2	19
3	30
4	38
5	50
6	61
7	70
8	79
9	88
10	98
11	110

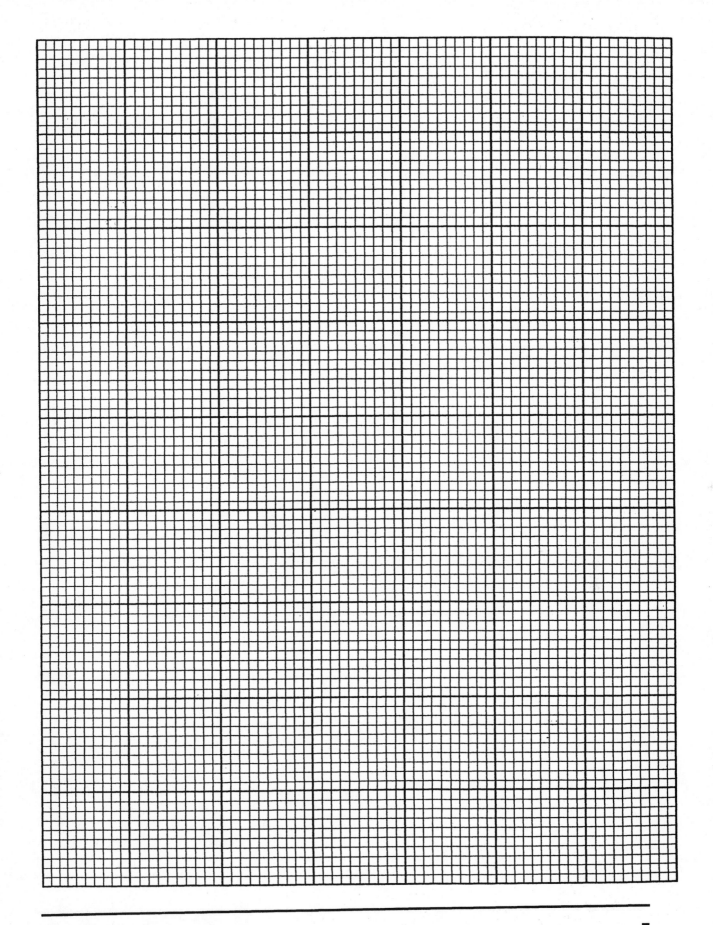

Two-Dimensional Graphing

EXPERIMENT 2

UNITS, ESTIMATES, ERRORS

Name_____ Partner(s)_____

Date_____ _____

Section Number_____ _____

BASE CONCEPTS: Measurement Units, Metric System, Mass

INTRODUCTION:

A measurement unit is an essential piece of information for any physical quantity. For instance, if someone asked you how far you drove today, the answer "10" would make no sense. Some units of length such as feet, meters, miles, or kilometers would also be needed in the answer. An answer to almost any physical problem is **not complete** if the units are omitted.

Almost all countries in the world (except Myanmar and the United States) use the metric system of units. In this system the meter is the fundamental unit of length, the kilogram is the fundamental unit of mass, the second is the fundamental unit of time, and the Celsius degree is the fundamental unit of temperature. By comparison, in the United States the foot is the fundamental unit of length, the pound is the fundamental unit of weight, the second is the fundamental unit of time, and the Fahrenheit degree is the fundamental unit of temperature.

INITIAL QUESTIONS:

A.   What would be three advantages of switching to the metric system? Explain why you think they are advantages.

B.	Write down at least one disadvantage of switching to the metric system. Indicate why you think this is a disadvantage.

Essentially all scientific texts use metric units. This lab manual is no exception. However, to help you begin your personal conversion process, some common (approximate) equivalences are listed next.

1 meter = 3.3 ft = 39 inches
1 cm = 0.39 inch
1 km = 0.62 mile
1 ft = 30 cm
1 in. = 2.5 cm
1 mile = 1.6 km
1 kg weighs 2.2 lb on Earth
1 lb on Earth has a mass of 0.45 kg
Celsius temperature = 5 × (Fahrenheit temperature − 32°)/9
Fahrenheit temperature = 32° + (9 × (Celsius temperature))/5

C.	A typical adult in the United States weighs 150 lb. This person's mass is _____ kg.

D.	Your height in U. S. units is _____, which is _____ cm.

E.	The distance to _____ from _____ is
	name of a city name of the city you are in

	roughly _____ (in U.S. units) and _____ (in

	metric units).

F.	A normal driving speed in town is _____(U.S.) and _____(metric).

The ability to estimate quantities is important in physics as well as in many other fields. This process is studied in more detail in Experiment #4 (*Physical Estimation*), but its use as a tool is introduced here. Estimation consists of quickly making an educated guess. Some people are better at estimating than others, but it is possible to improve with practice. Even poor estimators can do rather well when they focus on the task and use a system of units that is familiar to them.

To evaluate your estimating skills, you need to know how to compare your estimated value with a carefully measured one. To calculate the percent deviation (or "error"), find the positive difference between your estimate and the accepted value, divide by the accepted value, and multiply by 100%. In equation form this looks like

$$\%DEVIATION = \frac{|ESTIMATED\ VALUE\ -\ ACCEPTED\ VALUE|}{ACCEPTED\ VALUE} \times 100\%,$$

where the straight vertical lines (absolute value sign) indicate that you take the positive value of the quantity between them. Note that the estimated value and accepted value must be expressed in the same units.

PROCEDURE AND QUESTIONS:

1. Pick four objects around you (e.g., a doorway, a table) and estimate their length in U.S. units. Then, measure their length in the same units. Record your values in the following table, and calculate the percent deviation for each estimate. Be sure to include the units in which you estimated and measured.

OBJECT	ESTIMATED VALUE	MEASURED VALUE	% DEVIATION

2. Now choose four *other* objects and estimate their length in metric units. Try to estimate directly in metric units rather than converting from U.S. units. Again measure the lengths and calculate the percent deviations. Do your best.

OBJECT	ESTIMATED VALUE	MEASURED VALUE	% DEVIATION

➤G. Based on these results, estimate the percent error you would expect when visually estimating the length of common objects.

➤H. Explain the reasoning that led to your percent error estimate in G.

➤I. Did you do better in U.S. or in metric units?

➤J. Why do you think your results came out the way they did?

Estimation is not limited to one value at a time. You can estimate a wide range of quantities with limited knowledge. The next set of questions asks you to estimate a variety of different values. Some results you can check directly, as you did above; others you cannot.

Units, Estimates, Errors

3.   Make estimates and explain the *process* you used in estimating for each of the following problems. Some reasonable ways to start your estimation might include estimating lengths to obtain areas and volumes; estimating densities by roughly comparing them to the density of water which is 1 gram/cubic centimeter; or counting in a small region and multiplying up to a larger one.

➤K.   Estimate the height of this building. (*Hint:* How many stories in the building? What is the approximate height per story?)

➤L.   Estimate the volume of this room. (Volume = height × length × width.)

➤M.   Estimate the total mass (in kilograms) of the air in this room. (Air is about 1000 times less dense than water.)

➣N.   Estimate your volume.

➣O.   Estimate your density (mass/volume).

➣P.   Estimate the number of houses in a typical square block in your town.

➣Q.   Estimate the number of seconds in 10 years (no calculators please). Make sure you explain **how** you arrived at your estimate.

GLOBAL QUESTIONS:

R.   When calculating the percent error, why do you need to make sure that the estimated and measured values are expressed in the same units?

S.   Which of the preceding estimates (K– Q) do you think you could easily verify? Why?

T.   Verify (by direct measurement) one of the estimates noted in Question S and find the percent deviation between your estimated and the measured values.

U.   State three common processes that would be made simpler by estimating first. Explain why the estimation would make the process simpler. An example (one that you unfortunately may not use as part of your answer) is preparing for a party. An estimation of the number of invited guests who will actually show up makes the process simpler because the person holding the party can then estimate the amount of food to buy and the size of the room needed for the party.

V.   Do you think that you must be able (eventually) to know the "correct" answer for estimation to be useful? Why or why not?

EXPERIMENT
3   HOW FAR TO A STAR?

Name_____   Partner(s)_____

Date_____   _____

Section Number_____   _____

BASE CONCEPTS: Arithmetic Operations, Speed

INTRODUCTION:

This lab introduces you to physics problems that require numerical calculations and to the use of numerical methods in studying some interesting astronomy questions. To help you develop confidence in your calculations, answers to some of the questions are given at the end of this exercise. Please *try the problems first*, before looking at the answers.

PROCEDURE AND QUESTIONS:

1. **SCIENTIFIC NOTATION**

   Scientific notation is a convenient way to write very large or very small numbers in condensed form. This makes calculations easier. The general approach is to make the number more compact by moving the decimal point (thus eliminating zeros that are merely place holders). Then, to restore the correct magnitude, the number is multiplied by 10 raised to a power equal to the number of places the decimal point was moved. When the decimal point is moved to the left, the exponent is positive; to the right, it is negative. Here are two examples: 10 can be written as 1.0×10^1 (clearly, a decrease in compactness, but a good illustration), and 0.005 can be written as 5×10^{-3}. Many calculators express large or small numbers in scientific notation.

➤A.   Write the following numbers in scientific notation.

     A1.   45100 =

     A2.   310,000,000,000 =

     A3.   0.001000 =

     A4.   0.0000000000012 =

When multiplying two numbers written in scientific notation, add the powers of 10. When dividing, subtract the power in the denominator from the power in the numerator. For example, if you multiply 300 by 20, without using scientific notation, it is clear that $300 \times 20 = 6000$. In scientific notation, $(3 \times 10^2)(2 \times 10^1) = 6 \times 10^{(2+1)} = 6 \times 10^3 = 6000$! If you divide 400 by 40, you get 10. In scientific notation, you have $(4 \times 10^2)/(4 \times 10^1) = 4/4 \times 10^{(2-1)} = 1 \times 10^1 = 10$, as before. These examples are fairly trivial, and under normal circumstances you wouldn't bother to convert these numbers to scientific notation. Nonetheless, they clearly illustrate the processes involved. Now it's your turn to try.

➤B. Evaluate the following products and quotients using scientific notation.

   B1. $(2 \times 10^3) \times (4 \times 10^7) =$

   B2. $(10^9) \times (2 \times 10^{-2}) =$

   B3. $(4 \times 10^7)/(2 \times 10^5) =$

   B4. $(4 \times 10^7)/(2 \times 10^{10}) =$

➤C. What is the smallest "large number" that your calculator expresses in scientific notation when you enter it without using scientific notation?

➤D. What is the largest "small number" that your calculator expresses in scientific notation when you enter it without using scientific notation?

2.   NUMERICAL ESTIMATES

When doing a complex calculation, it is often helpful to make a quick estimate of the answer. Making a quick, rough calculation in your head is useful in many situations (e.g., you want to know quickly if the restaurant check for dinner is approximately correct). A good way to get a rough idea of the answer is to write down (or visualize) the numbers in scientific notation rounded to just one digit (or, at most, two digits). For example, you want to know roughly the value of $408 \times 380/16400$. You can rewrite this as $(4 \times 10^2)(4 \times 10^2)/(16 \times 10^3)$ $= 16 \times 10^4/(16 \times 10^3) = 10$. This is not exact, but it is a fairly good approximation and much faster than calculating the exact answer.

➤E. To get some experience with such estimates, answer the following questions.

E1. What is the exact answer in the example 408 × 380/16400? Compare it with the estimated value.

First, estimate the answers for each of the following calculations, then use a calculator to compute the exact values to see how good your estimates are.

E2. 314 × 4021 =

E3. 4021 × 2893/113 =

E4. 10,257,936/(518 × 2039) =

E5. 0.0001629/84913 =

3.   **UNITS AND UNIT CONVERSIONS**

Most physical quantities must be expressed with both a number and a measurement unit. For example, suppose it is *three miles* to your favorite pizza shop. The word *mile* specifies the measurement unit.

To convert a quantity from one unit to another, you must use an appropriate *conversion factor*. For example, a high jumper clears 5.7 feet, and you want to know what this is in inches. In order to convert it you must know how many inches there are in a foot. It is helpful to write this information as a "conversion equation."

➤F.   To get used to conversion equations, answer the following question.

   F1.   1 foot = _____ inches.

This equation says that the left side and the right side are equal. Thus, if the right-hand side is divided by the left-hand side, that ratio equals 1 (a "pure number" with no units). Since you can always multiply any quantity by 1 and get the same quantity back, you can use the conversion equation to convert units by multiplying by 1. In Question F1, 1 foot = 12 inches. By dividing both sides by 1 foot, the equation becomes 12 inches/1 foot = 1 foot/1 foot=1. Therefore, 5.7 feet = 5.7 feet × 1 = 5.7 feet × (12 inches/1 foot) = 68.4 inches. Units cancel in the same way numbers do. If there is a unit "foot" in both the numerator and denominator of a fraction, they cancel each other. If you multiply by the "wrong 1," you will have a number that is correct, but it probably won't be useful. Using the same example, 5.7 feet = 5.7 feet × 1 = 5.7 feet × (1 foot/12 inches) = 0.48 feet × feet/inch. This is a "correct" value, but instead of units of inches, it has units of feet squared/inch.

➤G.   Convert each of the following. If you do not know the conversion equation, look it up in one of the books in the lab. State the conversion equation you used and show how you used it.

   G1.   2640 feet = _____ miles

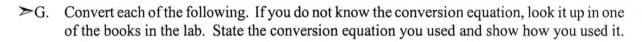

   G2.   3.2 hours = _____ minutes

G3.   200 kilometers = _____ miles

G4.   500 miles = _____ kilometers

4.   **HOW FAR TO A STAR?**

Now you can use the techniques you learned in Procedures 1, 2, and 3 to explore some questions about astronomy and the speed of light. Show all your work on these pages. Be sure to include units. Estimations are appropriate; use them. They will save you a lot of work and keep you from getting bogged down in unnecessary details. For some of these questions, you will need to recall that speed is distance divided by time (or distance = speed × time). **In all cases, show your work.**

➤H.   It is about a quarter of a million miles to the Moon (2.5×10^5 miles). The speed of light is 186,000 miles/second (about 2×10^5 mi/s). How much time will it take a laser beam to reach the Moon?

➤I.   What is the distance to the Moon in kilometers (km)?

➤J. It takes light about 8 minutes to travel here from the sun. What is the distance from the sun to the earth in miles?

➤K. How long would it take for a radio wave to get from here to China (halfway *around* the earth)? The radius of the earth is approximately 4000 miles. [Note that radio waves are related to light and travel at the same speed]

➤L. If the entire sun were suddenly turned off, we would not become aware of this fact until 8 minutes later. At that time, we would see a small dark spot expanding from the center of the sun (as seen in the sky) until the entire sun became dark. Why would we see this spread of darkness, rather than a sudden darkening of the sun?

➤M. If we look at the sun in the sky (never do this with unshielded eyes), we will see what looks like a flat disk. Because the sun is a sphere, the center of the disk that we see is really about half a million miles closer to us than the edge. Use this information to calculate how long the spread of darkness in Question L would take. (*Hint*: You don't need to use the distance from the sun to the earth to answer this.)

➤N. You have been told that light travels here from the sun in 8 minutes. The distance to the sun from the earth is thus said to be "8 light-minutes." That is, a *light-minute* is the distance light travels in 1 minute and is a unit of distance. The distance between the sun and Mars (radius of Mars' orbit) is 13 light-minutes. How close can Earth and Mars approach each other (in light-minutes)? (*Hint*: Draw a diagram showing the sun and the orbits of Earth and Mars; also indicate the radii of the two orbits as measured in light-minutes.)

➤O. How far apart can Earth and Mars get (in light-minutes)?

➤P. An *astronomical unit* (AU) is a unit of length defined to be the distance from Earth to the sun (8 light-minutes). A *light-year* (LY) is the distance light travels in one year. How many AU are there in 1 LY? (*Hint*: You know that 1 AU = 8 light-minutes, and you can convert light-minutes to light-years.)

➤Q. Our nearest neighboring star (this is not the sun) is 4 LY away. How far is this in AU?

➤R. If your speed is 1/20 the speed of light, how long will it take you to reach this nearest star?

➤S. The *Pioneer* space vehicle travels at about 25,000 mph. How long does it take *Pioneer* to travel 1 AU? Feel free to use answers you have already obtained.

➢T. Pluto, the outermost planet in our solar system, travels in an orbit whose radius is about 40 AU, so it is about 40 AU away from the sun. Thus, the distance from Earth to the edge of the solar system is about 39 AU. How long will it take *Pioneer* (Question S) to travel to the edge of the solar system? Feel free to use your answer from Question S.

➢U. How long would it take *Pioneer* to travel to the nearest neighboring star mentioned in Question Q?

➢V. The Crab Nebula is the remnant of a supernova explosion. It is about 3500 LY away. The Chinese *saw* this star explosion in 1054 ad. In what Earth year did the star *actually* explode?

ANSWERS TO SELECTED QUESTIONS FROM PROCEDURES 1, 2, AND 3

A1. 4.51×10^4

A3. 1.000×10^{-3}

B1. 8×10^{10}

B3. $2 \times 10^2 = 200$

E1. 9.45

E3. 1.03×10^5

E5. 1.92×10^{-9}

G1. 0.5 mile

G3. 124 miles

EXPERIMENT 4

PHYSICAL ESTIMATION

Name_____   Partner(s)_____

Date_____   _____

Section Number_____   _____

BASE CONCEPTS: Measurement, Estimation, Percent Error, Mass

INTRODUCTION:

Estimation is the process of making an educated guess. The uses of successfully estimating unknown quantities are many and varied. For example, if you want to distribute pamphlets to people in your community, it is necessary to know about how many pamphlets are required. The easiest and most practical approach is to estimate the number of households, actually counting them would normally be difficult, costly, and time consuming. If you are planning a meeting, you need to estimate the number of people attending so that you can reserve a room of the correct size, order the right amount of coffee, and so on. In the sciences, estimating an answer is an important way to check a calculation or to get an approximate value for a quantity that you cannot (readily) calculate exactly. This lab gives you a start on understanding how to make estimates and appreciating some factors that contribute to accurate estimation.

In this lab you will estimate different physical quantities (length, time, and mass). At the end of the lab, you will be asked questions comparing the processes you used to make these estimates.

SUPPLIES:

Several sheets of paper of various sizes, metric ruler, timer, scale, various masses.

PROCEDURE AND QUESTIONS:

1. Take one of the pieces of paper, and using a book as a straightedge, draw what you think is a 15-centimeters line. Be careful to clearly mark the end points.

➤A.   Why did you draw the line the length you did? That is, what guided you in making this estimation?

2.   Now, use a metric ruler to measure the actual length of the line that you drew as an estimation of 15 cm. (You should measure to the nearest 0.1 cm = 1 mm.) This line is actually _____ cm long.

3.   On the *same* piece of paper, again using your book as a straightedge, draw a line that you estimate to be 9 cm long.

➤B.   What guided you in making this estimation? How was this estimation similar to and/or different from the first one?

4.   Measure the actual length (to 1 mm). The line that you estimated to be 9 cm is actually _____ cm long.

5.   On a *different* piece a paper, estimate a 10-cm line.

➤C.   What guided you in making this estimation? How was this estimation similar to and/or different from the first two?

6.   Measure the actual length. The line that you estimated to be 10 cm is actually _____ cm.

7.   On *this second piece of paper* draw an estimated 12 cm line.

➤D.   What guided you in making this estimation? How was this estimation similar to and/or different from the first three?

8.   Measure the actual length. The line that you estimated to be 12 cm is actually _____ cm long.

➤E.   Which of these four estimates do you think is the most accurate?

➤F.   What definition of accurate did you use to answer Question E.

A quantitative way to specify the accuracy of an estimate is in terms of <u>percent deviation</u> (or percent error, or relative error). To calculate it, you compute the *positive difference* between the actual (measured) value and the target value, divide by the target value, and multiply by 100%. For example, in the first case the target length was 15 cm. If the line you drew was actually 10.5 cm long, then the positive difference is 15.0 cm − 10.5 cm = 4.5 cm. The percent error is then $\frac{4.5 \text{ cm}}{15.0 \text{ cm}} \times 100\% = 0.30 \times 100\% = 30\%$. Note that a **smaller** percent error corresponds to a **more accurate** estimate, and a larger one to a less accurate estimate. The actual value and the target value must have the same units. This ensures that the physical units cancel in the calculation, and the result is a "pure" percentage.

9. Complete the following table after calculating the requested values from your findings. Rank your results 1 through 4, with 1 being the most accurate and 4 being the least accurate.

TARGET	ACTUAL (Measured)	Δ (Positive Difference)	$\dfrac{\Delta}{\text{TARGET}} \times 100\%$	RANK
15 cm				
9 cm				
10 cm				
12 cm				

➤G. Explain what factors you think caused your accuracies to come out the way they did. That is, why do you think you estimated the more accurate ones better than the less accurate ones?

10. Tabulate your rankings on the board. After all the groups have put their results on the board, your instructor will add them up and give you values to enter in the following table.

Number of people in this lab: _____

TARGET VALUE	Number of People Who Ranked Target As			
	1	2	3	4
15 cm				
9 cm				
10 cm				
12 cm				

➤H.   Is there a trend for the class? If so, describe the trend.

➤I.   Why do you think this trend occurred?

11.   Now you are going to estimate the time it takes for one minute to pass. Work in groups of at least two with each group having a timer. One person at a time, each member of the group should close his or her eyes and, at the same time, another member of the group should start the timer. The person whose eyes are closed is to estimate the time it takes 1 minute to pass by telling the timekeeper to stop after an estimated minute.

The actual elapsed time during your estimated minute was _____ seconds.

The percent error for this estimation was _____ %.

12.   After all members of your group have estimated 1 minute, each member should repeat the exercise, this time estimating 30 seconds.

The actual elapsed time during your estimated 30 seconds was _____ seconds.

The percent error for this estimation was _____ %.

➤J.   In which trial were you more accurate?

13.   Each person in the group should again estimate 1 minute.

The elapsed time for this estimated minute was _____ seconds.

The percent error for this estimated minute was _____ %.

➤K.   Did estimating one minute previously help you in estimating 1 minute this time?   Why or why not?

➤L.   How is this process the same as and/or different from estimating length?

14.   You are now going to estimate the masses of various objects. Pick up a block of wood and estimate its mass in grams.

The estimated mass of the wood block is _____ g.

15.   Now, measure its mass with a scale.

The actual mass of the wood block is _____ g.

Physical Estimation

➢M. Explain why you made the estimation you did for the wood block's mass.

16. Repeat Procedures 14 and 15, using a metal rod instead of a block of wood.

The estimated mass of the metal rod is _____ g.

The metal rod's actual mass is _____ g.

➢N. Explain why you made the estimate you did for the rod's mass.

17. Finally, take a bag of Styrofoam "peanuts" and estimate its mass.

The estimated mass of the peanuts is _____ g.

The actual mass of the peanuts is _____ g.

➢O. Explain why you made the estimate you did for the mass of the peanuts.

18. Compute the percent difference between the actual and estimated masses for each of the three objects.

19. List those estimates that were larger than the actual masses.

List those estimates that were smaller than the actual masses.

20. On the board, order your estimates of the wood, Styrofoam, and metal masses 1 through 3, with 1 being the most accurate of the three, and 3 being the least accurate. When placing this information on the board, also use a + if your estimate was higher than the actual value and a − if your estimate was lower. For example, if your worst accuracy was for the wood block for which you estimated 300 g when its actual mass was 200 g, you would put +3 in the column for wood.

Look at the board after all the lab participants have placed their information on it. Answer the following questions *based on the class information*.

➤P. Is there any general trend for the class with respect to which measurements were the most or least accurate? If so, give some reasons why this trend might exist.

➤Q. For the class, do you see any trend for which the estimated measurements were high (+) or low (−)? If so, give some reasons why this trend might exist.

GLOBAL QUESTIONS:

The following questions refer to all three types of estimation: length, time, and mass.

R. In each of the three types of estimation, you had some self-correcting technique. How were these three techniques similar?

S. How were the techniques different?

T. For which type of estimation was the self-correction most effective? Why?

U.   For which type of estimation was the self-correction least effective? Why?

V.   In what ways do you think you could improve your abilities to estimate time, length, and mass?

EXPERIMENT 5

MEASUREMENT UNCERTAINTY

Name_____   Partner(s)_____

Date_____   _____

Section Number_____   _____

BASE CONCEPTS: Accuracy, Estimation

INTRODUCTION:

This laboratory has two distinct parts. In the first part, you will measure the length of your lab table in different ways and compare the accuracies you achieve. In the second, you will measure the period of a pendulum several times to see how the resulting accuracy is affected by the number of measurements performed.

SUPPLIES:

Lab table, meterstick, 2-meter stick or steel tape, pendulum, timer.

PROCEDURE AND QUESTIONS:

1.   By looking at the lab table, estimate its length (the longer of the two horizontal dimensions) in centimeters as accurately as you can. Before going on to Procedure 2, write down an indication of how accurate you think this estimate is (e.g., write down ± 50 cm if you are confident that your value is within ± 50 cm of the "true" length, but you are not confident that it is within ± 40 cm). This is your estimate of the uncertainty in your length estimate. (Don't forget the units!)

My estimated length is _____.

My estimate of the uncertainty is _____.

2.   Carefully measure the length of the table using a meterstick. Measure to at least the nearest millimeter (mm). Thus, if your result is written in centimeters, you will have at least one digit to the right of the decimal point. Estimate the uncertainty in this measurement.

My measured length is _____.

My estimated measurement uncertainty is _____.

3.   Now, carefully measure the length of the table using a 2-meter stick or steel tape. Again measure to *at least* the nearest millimeter. Estimate the uncertainty in this measurement.

My measured length is _____.

My estimated measurement uncertainty is _____.

➢A.   Which of the three length values is the most accurate? On what do you base your answer?

➢B.   How could you decrease the uncertainty in the best measurement of the lab table's length?

➢C.   Could you ever measure the table with no uncertainty?   Why or why not?

When performing experimental measurements, there are three basic types of "errors" that one must consider:

Mistake   These are errors that are due to the experimenters themselves and **not** due to the experimental procedures. They can be eliminated from experiments by the exercise of proper care.

Systematic   These are errors in an experiment that cause the results to be "off" in the *same direction* every time the experiment is performed. As an example, suppose that you are using a poorly made timer to measure your pulse rate. If the timer always runs slow, you will *always* measure a value that is higher than your actual pulse rate.

Random   These are errors in an experiment that cause the results to be "off" in *either direction* (sometimes high, sometimes low) in an essentially random way.

➤D.   In the experiment of measuring the lab table, what is one possible source of each of the three types of errors?

Mistake:

Systematic:

Random:

➤E.   If you took many measurements and averaged them, which one of the three types of errors would be reduced? Why?

In this part of the lab, you are to measure the period of a pendulum. The period of a pendulum is defined as the time it takes the pendulum to complete one full cycle, that is, the time to swing back and forth once so that it returns to its initial position moving in its initial direction. This time should depend only on the shape of the pendulum and its length. The lengths of the pendulums have all been set so that the periods should be the same. **Please do not change the length of your pendulum.**

4.   In groups of two or three, take 45 separate measurements of the period of your pendulum. Enter your measurements in Table 1.

Enter the values of the 45 measurements in 45 rows of one column of a spreadsheet or as a list in your calculator.

Table 1
Measurements of Periods

TRIAL #	PERIOD (s)	TRIAL #	PERIOD (s)	TRIAL #	PERIOD (s)
1		16		31	
2		17		32	
3		18		33	
4		19		34	
5		20		35	
6		21		36	
7		22		37	
8		23		38	
9		24		39	
10		25		40	
11		26		41	
12		27		42	
13		28		43	
14		29		44	
15		30		45	

Find the sum and the average or mean (the sum divided by the number of trials) of the 45 trials.

The **sum** of the 45 trials is _____ seconds.

The **average** (to the nearest 0.01 second) period is _____ seconds.

5. Sort the 45 measured periods from smallest to largest, and if using a spreadsheet, print out the sorted list. If not using a spreadsheet, write the ordered list on a piece of paper.

6. Circle the average time on your sorted list. There may not be a value on the list that exactly matches your average. In this case, circle the two values that fall on either side of the average. If there are several experimental values equal to the average, circle only one of them.

Table 2
Example of Sorted Data

PERIOD (s)	PERIOD (s)	PERIOD (s)	PERIOD (s)	PERIOD (s)
2.09 (smallest)	2.17	2.19	2.20	2.22
2.12	2.17	2.19	2.20	2.22
2.15	2.18	2.19 (average)	2.20	2.22
2.15	2.18	2.19	2.20	2.24
2.16	2.18	2.19	2.21	2.24
2.16	2.18	2.19	2.21	2.25
2.16 (smallest 30)	2.18	2.20	2.21	2.25
2.17	2.18	2.20	2.21	2.29
2.17	2.19	2.20	2.21 (largest 30)	2.29 (largest)

7. Find the 30 values that are closest to the average. They may be either higher or lower than the average. Only the magnitude of the difference between the value and the average time matters. Be careful. You cannot simply count the first 15 values below and the first 15 above the average. It is possible that there will be more values close to the average in one direction than in the other. See Table 2 on p. 41 for an example. Enter the largest value that was within the 30 closest and the smallest value that was within the thirty closest.

The **largest** value in the 30 closest is _____ seconds.

The **smallest** value in the 30 closest is _____ seconds.

8. Tally your data on the board along with data from all the other lab groups. The total number of groups contributing their data for this tally is _____. This number will be called N_g in the discussion that follows.

9. Repeat Procedures 6 and 7 for the data from the entire class. This time, however, instead of finding the closest 30 values to the average, find the closest $30 \times N_g$ values.

The **average** period for the entire class is _____ seconds.

The **largest** value in the $30 \times N_g$ closest is _____ seconds.

The **smallest** value in the $30 \times N_g$ closest is _____ seconds.

➤F. How does the average period differ in Procedures 6 and 9?

➤G.   How do the largest and smallest periods differ in Procedures 7 and 9?

➤H.   The collection of more data helps in obtaining "THE" value. How?

➤I. Describe a better (as in faster, more accurate, easier, etc.) way to determine the period of a pendulum.

Measurement Uncertainty

EXPERIMENT 6

MEASUREMENT AND STATISTICS

Name_____   Partner(s)_____

Date_____   _____

Section Number_____   _____

BASE CONCEPTS: Percentage Difference, Histograms

INTRODUCTION:

The discipline of statistics is often praised as an indispensable means of extracting information from chaos and confusion. It is equally often derided as an obscure tool for deceit and dishonesty. With the advent of computers, the use of statistics has become increasingly widespread, directly or indirectly influencing almost every phase of contemporary life. In this laboratory you will explore (at an elementary and intuitive level) the essence of statistics.

SUPPLIES: Coins, ruler.

PROCEDURE AND QUESTIONS:

➤A. If you toss six coins in the air and count the number of heads up when they land, how many heads up do you expect there will be?

Why?

1. Toss six coins at once. Count the number of heads that come up on the six coins and enter a tick mark by the appropriate number in Column 2 of Table 1 on the following page.

2. Repeat Procedure 1 **at least** 35 more times. Since you will want to count the number of tick marks for each possible outcome, a convenient way to make the marks is to make four vertical ticks and then a fifth diagonally over the first four. Total the number of tick marks for each outcome in the third column of Table 1. Because you will need this information later, fill in Column 4 of Table 1 by multiplying the number in Column 1 by the number in Column 3.

The number of times (N_m) you threw the six coins (**number of trials**) is _____.

Measurement and Statistics   **45**

Table 1
Group data

COLUMN 1 # OF HEADS	COLUMN 2 TICK MARKS	COLUMN 3 # OF TICKS	COLUMN 4 # HEADS × # TICKS
0			
1			
2			
3			
4			
5			
6			

3. On the "class total board," enter the number of tick marks you have for each possible outcome (number of heads). Your instructor will total the results from all the groups and enter these totals on the board. Copy the results for the entire class in Column 2 of Table 2 or in a spreadsheet, and calculate the percentage of times each of the seven outcomes occurred. (This is the number of times the outcome occurred (Column 2) divided by the total number of trials, N_e, multiplied by 100%.) Again, complete Column 4 for later use. Refer to Procedure 2 if necessary.

The total number of trials for entire class (N_e) is _____.

Table 2
Entire class' data

COLUMN 1 # OF HEADS	COLUMN 2 TICK MARKS	COLUMN 3 PERCENTAGE	COLUMN 4 # HEADS × # TICKS
0			
1			
2			
3			
4			
5			
6			

4.   Use a spreadsheet or your ruler to draw two bar graphs or histograms, showing the total number of times each outcome occurred, versus the number of heads (0, 1, 2, 3, 4, 5, 6). The first histogram should be for your own data, and the second should be for the combined data from the entire class. Examples of typical data and the graphing of the data can be found in the following sample data.

SAMPLE DATA, CALCULATIONS, AND GRAPH

#HEADS	# TICKS	PERCENT AGE	# HEADS × # TICKS
0	0	0%	0×0=0
1	1	10%	1×1=1
2	2	20%	2×2=2
3	6	60%	3×6=18
4	1	10%	4×1=4
5	0	0%	5×0=0
6	0	0%	6×0=0

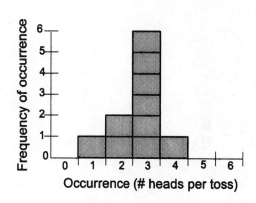

Number of trials = sum of Column 2 =
$0 + 1 + 2 + 6 + 1 + 0 + 0 = 10$

Mean = sum of Column 4/number of trials = $(0 + 1 + 2 + 18 + 4 + 0 + 0)/10 = 2.5$

Number of occurrences on either side of 2.5 needed to include *at least* 2/3 (7) of the trials = 1

Standard deviation = 1

Standard error = $1/\sqrt{10} = 0.32$

Measurement = 2.5 ± 0.32

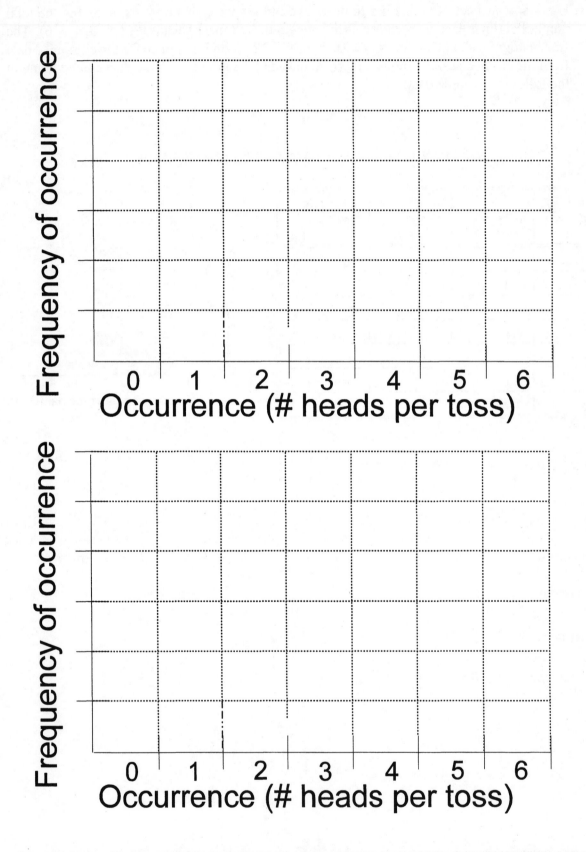

➤B.   How are the total class results different from the results using your data alone?

How are they the same?

➤C.   Are these results surprising to you?

Why or why not?

➤D.   What do you think the results would look like if you had 10 times as many tosses on your graph?

Why?

➤E.   Why is it important to have "lots of data"?

---

5. To compute the mean for **your data** you need to multiply the number of heads for each outcome (0, 1, 2, 3, 4, 5, 6) by the number of times that outcome occurred. Then, add all seven of these calculated values together, and, finally, divide by the total number of data points. You should have done the first of these steps and entered the result in Column 4 of Table 1.

Number of trials (tosses) for my group (N_m) (see Procedure 2 if necessary) = _____ .

Sum of Column 4 in Table 1 is _____ .

Sum of Column 4 divided by number of trials (N_m) _____ = mean of my data.

6. Now complete the mean for the entire class's data using the same set of procedures.

Number of trials (tosses) for entire class (N_e) (see Procedure 3 if necessary) = _____ .

Sum of Column 4 in Table 2 is _____ .

Sum of Column 4 divided by number of trials (N_m) _____ = mean of class's data.

➤F. Which of these more closely matches the value you initially expected to throw? (See your answer to Question A for help.)

Give a reason to justify this result.

The remainder of the lab is intended to enable you to develop some feel for the statistical concept of *standard deviation* and *standard error*.

When experimental scientists present their results, they never present just a numerical value. Rather, they present a value for the measured quantity plus or minus a second number. The first value is usually the mean of their individual measurements (or data points). The second number is the statistical uncertainty (or error) of the mean. It is obtained by dividing the standard deviation of the data points by the square root of the number of measurements. The standard deviation is calculated from a straightforward but somewhat complicated mathematical formula. Its *significance*, however, is rather simple. For normally distributed data (data that fall nicely under a bell-shaped curve), just over two-thirds of the data should fall within one standard deviation of the mean (on either side). Thus, for a result obtained from N data points and presented as $\mathbf{A} \pm \mathbf{B}$, a little more than two-thirds of the data points should fall between $\mathbf{A} - \mathbf{B} * \sqrt{N}$ and $\mathbf{A} + \mathbf{B} * \sqrt{N}$. As an example, suppose an experiment has been performed to determine the length of a meterstick. The result, based on 25 measurements, is presented as 1.002 ± 0.005 m. This statement tells you that the mean of the 25 measurements is 1.002 m and that the standard deviation is 0.005 m $\times \sqrt{25} = 0.025$. This tells you that about 17 measurements [17 = (2/3) x 25] fell between the mean minus the standard deviation (0.977 m = 1.002 m - 0.025 m) and the mean plus the standard deviation (1.027 m = 1.002 m + 0.025 m). In the following procedures, you will investigate how the standard deviation and statistical uncertainty change with the number (N) of data points.

7.   Clearly mark the mean on the histogram for **your data**. Find the number of bins you need to include symmetrically on each side so *at least* two-thirds of your trials fall in your included bins. See the example calculations if needed.

The mean for your group's data is _____.

The number of data points, N_m, is _____.

Two-thirds (2/3) of the number of data points is _____.

The standard deviation (σ_m) (number of bins on either side of the bin that includes the mean so that at least two-thirds of the trials are included) is approximately _____.

The standard error ($\sigma_m/\sqrt{N_m}$) for your group's trials is _____.

8.   Now, repeat this process for the data from the **entire class**.

The mean for the entire class's data is _____.

The number of data points, N_e is _____.

Two-thirds (2/3) of the number of data points is _____.

The standard deviation (σ_e) is _____.

The standard error ($\sigma_e/\sqrt{N_e}$) for your group's trials is _____.

➤G. Which of the two standard deviations is lower — for your data or for the entire class?

Give a reason for this outcome.

➤H. Which of the two statistical errors is lower — for your data or for the entire class?

Give a reason for this outcome.

From this outcome explain why more data points are helpful in obtaining accurate results.

EXPERIMENT 7

EXPERIMENTAL DETERMINATION OF THE VALUE OF π BY DIRECT AND INDIRECT METHODS

Name_____   Partner(s)_____

Date_____   _____

Section Number_____   _____

BASE CONCEPTS: Mass, Geometry

INTRODUCTION:

Scientists do not always measure quantities they need to know in a direct way. If it is not possible to measure the desired quantity directly, or if the quantity can be determined more accurately by other means, they may instead measure one or more values related to the desired quantity. Using known relationships between the measured values and the desired quantity, they can compute the desired result. This procedure is called an *indirect* determination or measurement.

In this laboratory you will determine experimentally the value of pi (π), defined as the ratio of the circumference of a circle to its diameter, by both direct and indirect means.

The circumference of a circle is given by πd, where d is the diameter of the circle, and the area of a circle is given by $\pi d^2/4$. Also, there is a direct relationship between the mass of a uniform piece of paper and its area. If this relationship is known, the mass of a circular piece of paper can be used to compute the value of π. This technique provides an indirect determination of π.

SUPPLIES:

Cardboard, scissors, metric ruler, circular compass, scale, fishing line.

PROCEDURE AND QUESTIONS:

1.   In the first part of this laboratory, you will measure π *directly*. Draw a circle with a diameter of roughly 20 cm on a piece of cardboard. Carefully cut out the circle, then measure its diameter (in centimeters) as precisely as you can.

The diameter of this circle (CIRCLE #1) is _____.

➤A. What technique did you use to determine the diameter of the circle? Try to describe it as clearly as possible.

2. Carefully measure the circumference of the circle. **Do not** use the equation to calculate it!

   The measured circumference is _____.

➤B. What technique did you use to determine the circumference? Try to describe it as clearly as possible.

3. Calculate a value for π from your measurements (π = circumference/diameter).

   My direct value for π is _____.

4. Collect values of diameter, circumference and π from other lab groups and place them in a spreadsheet or in the following table.

Table 1
Group Data

DIAMETER	CIRCUMFERENCE	VALUE OF π

5. Calculate the average value of π by using all results from the class.

The average directly determined value for π is _____.

6. Now, you can find out how accurate this average value is. Given that the accepted value of π is 3.14159, compute the percent deviation (or percent error) for the average value. (*Remember:* Percent error = the positive difference between accepted value and average value, divided by accepted value, times100%.)

The percent deviation is _____.

7. The remainder of this lab consists of determining π *indirectly*. Obtain a rectangular piece of cardboard. Carefully measure its mass, length, and width. Multiply the length and width to obtain its area.

The mass of the entire piece of cardboard is _____ g.

The length is _____ cm, and the width is _____ cm.

The area of the entire piece of cardboard is _____ cm^2.

8.   Draw a circle about 20 cm in diameter on this cardboard. Carefully cut out the circle and measure its diameter.

The diameter of this circle (CIRCLE #2) is _____ cm.

➤C.   What technique did you use to determine the diameter?

9.   Now, measure the mass of this circular piece of cardboard.

The mass of CIRCLE #2 is _____ g.

10.   If the cardboard is uniform, the mass per area should be a constant. In particular, the mass of the entire sheet divided by its area should equal the mass of CIRCLE #2 divided by its area. Use your values for these quantities as determined above to calculate the area of CIRCLE #2:

$$\text{AREA of CIRCLE} = \text{AREA of SHEET} \times \frac{\text{MASS of CIRCLE}}{\text{MASS of SHEET}}.$$

The calculated value for the area of CIRCLE #2 is _____ cm^2.

11.   Since the area of a circle equals $\pi d^2/4$, where d is the diameter, π can be calculated by multiplying the area by $4/d^2$, that is,

$$\pi = \frac{4 \times \text{AREA of CIRCLE}}{\text{DIAMETER SQUARED}}.$$

My indirect value for π is _____.

12. Collect values of diameter, area, and π from other lab groups and place them in a spreadsheet or in the following table.

Table 2
Data from All Groups

DIAMETER	AREA	VALUE OF π

13. Compute the average value of π using all the data from the entire class.

The average indirectly determined value for π is _____.

14. Compute the percent deviation ("percent error") for the average obtained above (accepted value = 3.14159).

The percent deviation is _____.

GLOBAL QUESTIONS:

D. In both experiments, what was the desired value that you wanted to find?

E.   In the indirect measurement, what known relationship(s) did you use to compute the value of π?

F.   Which method "made more sense" to you?

Why?

G.   Looking at the entire class's results, which method do you think was more accurate?

Why do you think this was so?

H. Give three general reasons that a *direct* method might be used in an experiment.

I. Give three general reasons that an *indirect* method might be used in an experiment.

EXPERIMENT 8

KEPLER'S FIRST AND SECOND LAWS

Name_____   Partner(s)_____

Date_____   _____

Section Number_____   _____

BASE CONCEPTS: Plane Geometry, Kepler's Laws

INTRODUCTION:

An ellipse is a closed curve with the characteristic that the sum of the distances from any point on the curve to two specific fixed points inside the curve is a constant (see Figure 1). The two fixed points are called *foci* (plural of focus).

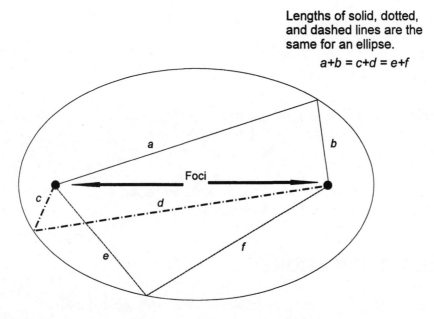

Lengths of solid, dotted, and dashed lines are the same for an ellipse.

$$a+b = c+d = e+f$$

Figure 1

You can easily construct an ellipse by tacking the ends of a string onto a sheet of paper, placing a pencil in the loop of the string, making sure that the string is pulled taut, and drawing completely around the curve (see Figure 2).

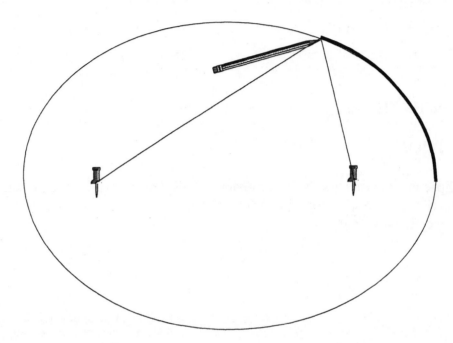

Figure 2

➤A. Indicate the locations of the foci in Figure 2.

SUPPLIES:

String, thumbtacks, metric ruler, scissors, pencil, unlined paper, tape, wallboard or cardboard, anything else you might want to use (but no compass).

PROCEDURE AND QUESTIONS:

1.   Use a piece of string 10 cm long to draw an ellipse with foci 5 cm apart.

The distance across this ellipse at its widest point (through its center) is _____.

The distance across this ellipse at its narrowest point is _____.

2. Using the *same string*, draw an ellipse with foci 8 cm apart.

   The distance across this ellipse at its widest point is _____.

   The distance across this ellipse at its narrowest point is _____.

3. Use this string to draw one more ellipse, this time with foci 0.5 cm apart.

   The distance across this ellipse at its widest point is _____.

   The distance across this ellipse at its narrowest point is _____.

➤B. How does the widest length change as the distance between the two foci increases?

➤C. How does the narrowest length change as the distance between the two foci increases?

➤D. A circle is an ellipse. Where are the foci?

4.   Draw a circle on a sheet of paper.

       For this circle, the constant sum of distances from the foci is _____.

       The radius of the circle is _____.

➤E. How is the constant sum of distances related to the radius of the circle?

➤F. Is a straight line of length 10 cm an ellipse? If so, where are the two foci? If not, why not?

Kepler's first law of planetary motion states that the planets move around the sun in elliptical paths with the sun at one focus. **Kepler's second law** states that a line joining the planet and the Sun sweeps out equal areas in equal times (see Figure 3). Qualitatively this means that the closer the planet is to the Sun, the faster it moves.

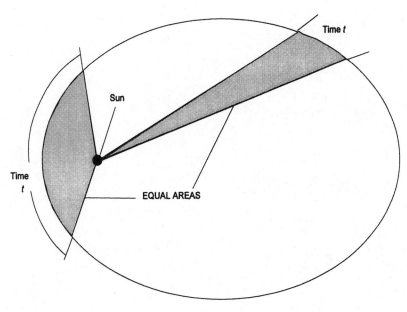

Figure 3

➤G. State in your own words why Kepler's second law implies that the closer a planet is to the sun, the faster it moves. Use pictures if it helps make your point.

(more writing space on next page)

5.   On the following page is a drawing (Figure 4) that represents a planet's orbit. The solid dot inside the ellipse signifies the focus at which the sun is located. The solid dots around the ellipse show locations of the planet at equal intervals of time. Divide your drawing into six groups of three time intervals. (See Figure 3 as an example.) For each of these groups, find the graphical area in square centimeters swept out by a line that connects the sun and the planet. One way to determine the area is to divide it into triangles and sum the areas of the triangles. The area of a triangle is found by multiplying its height by its base (these two are perpendicular to each other) and dividing by 2. There are other ways to determine the area, and you are free to use a different method. Think of techniques you may have used in other labs (e.g., Lab 4 and Lab 7).

Kepler's First and Second Laws

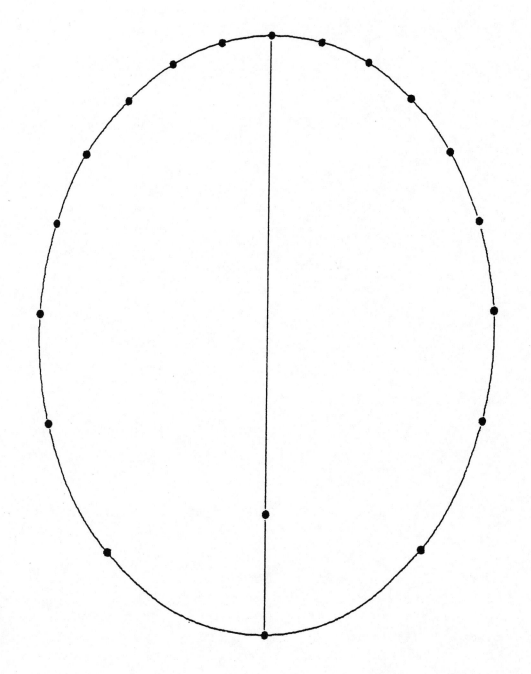

Figure 4

6.   Tabulate the areas of your six segments in a spreadsheet or in the following table.

SEGMENT	AREA (cm^2)
1	
2	
3	
4	
5	
6	

The sum of the six segment areas is _____.

The average segment area (sum divided by 6) is _____.

Determine the (maximum) percent deviation by analogy with the procedure used in studying estimation (see Lab 4). Treat the average segment area as your "target value" and the area that deviates farthest from the average as your "estimated value." The resulting equation then looks like:

$$\% \ \text{DEVIATION} = \frac{\text{AREA FARTHEST FROM AVERAGE} - \text{AVERAGE AREA}}{\text{AVERAGE AREA}} \times 100\%$$

The (maximum) percent deviation is _____.

GLOBAL QUESTIONS:

H.   If it takes 100 days for the planet to move from one point on your ellipse to an adjacent point, what is the period of this planet (the time required for the planet to go all the way around the sun)? Clearly explain your reasoning.

I.   If your ellipse models the orbit of the earth, how many days does it take the earth to move from one point to an adjacent point on the ellipse?

EXPERIMENT 9

THE SIMPLE PENDULUM

Name_____   Partner(s)_____

Date_____   _____

Section Number_____   _____

BASE CONCEPTS: Speed, Graphing (optional)

INTRODUCTION:

If an object is pulled back to its equilibrium position with an acceleration proportional to the distance from the equilibrium position, the object is said to exhibit *simple harmonic motion*. The time it takes such an object to return to some position, moving in the same direction with the same speed, is called the *period*. In simple harmonic motion, the period is independent of the maximum displacement from the equilibrium position (the amplitude of the motion).

In this lab you will study a pendulum that consists of a nonstretching string with a sphere of mass *m* at the end. The string is hung from a support as shown in Figure 1. If the angle the string makes with the vertical does not exceed 15°, the pendulum can be considered to exhibit simple harmonic motion. You will vary the initial angle the string makes with the vertical (the angular amplitude), the length of the pendulum, and the mass of the sphere independently to investigate how the period depends on each of these. The length of the pendulum is measured from the support point to the center of the sphere. The period is then the time it takes the pendulum to swing from one extreme to the other and back again.

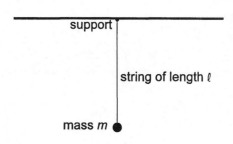

Figure 1

SUPPLIES: Pendulum bobs of various masses, pendulum support, protractor, meterstick.

PROCEDURE:

1.   Set the pendulum so that the length (from the top support to the center of the sphere) is 150 cm after measuring the mass of the sphere at the end. Record the mass of the sphere in a spreadsheet or in Tables 1 and 2 on p. 72.

---

2.   Using the protractor, start the pendulum with a 10° angular amplitude and measure the time for *ten periods* by starting the timer and turning it off after ten full oscillations. From this time, calculate the period of the pendulum and enter the data in a spreadsheet or Table 1.

3.   Repeat Procedure 2 for five different lengths, using the same sphere and making sure to start the pendulum at 10° to the vertical each time. Tabulate the collected data.

Table 1
Varying Length

Angular amplitude = 10°
Mass of sphere = _____ g

LENGTH (cm)	10 PERIODS (s)	1 PERIOD (s)

Table 2
Varying Amplitude

Length of pendulum = _____ cm
Mass of sphere = _____ g

ANGLE (degrees)	10 PERIODS (s)	1 PERIOD (s)

Table 3
Varying Mass

Angular amplitude = 10°
Length of Pendulum = _____ cm

MASS (g)	10 PERIODS (s)	1 PERIOD (s)

The Simple Pendulum

4. Choosing a pendulum length of about 100 cm, and continuing to use the same sphere, find the time for 10 periods with six different initial angles of your choice. Enter your data in a separate spreadsheet or in Table 2 on p. 72, including your chosen angles.

5. Using the same length as in Procedure 4, keeping the angular amplitude at 10°, vary the mass of the sphere, measure 10 periods, then calculate one period. Enter you data in a third spreadsheet or in Table 3 on p. 72.

GLOBAL QUESTIONS:

A. Does the value of the period change as the length of the pendulum changes? How do you know?

B. Does the value of the period changes as the mass of the sphere is changed? How do you know?

C. Does the value of the period changes as the amplitude changes? How do you know?

D. From an error (accuracy) point of view, why should you measure 10 periods and divide by 10 to get the period rather than just directly measuring a single period?

E.   If you took your pendulum to the moon, where the acceleration due to gravity is one sixth that on Earth, would you expect the period to be longer, shorter, or the same as on Earth? Why?

F.   Describe how a pendulum might be used to keep time.

G.   (Optional — talk to your instructor.) How does the period depend on the length of the pendulum? To investigate this, draw graphs of the period versus length, the period squared versus length, and the period versus length squared. On which one of these graphs do the data fall on a relatively straight line? What does this say about the relationship between the period and the length of the pendulum?

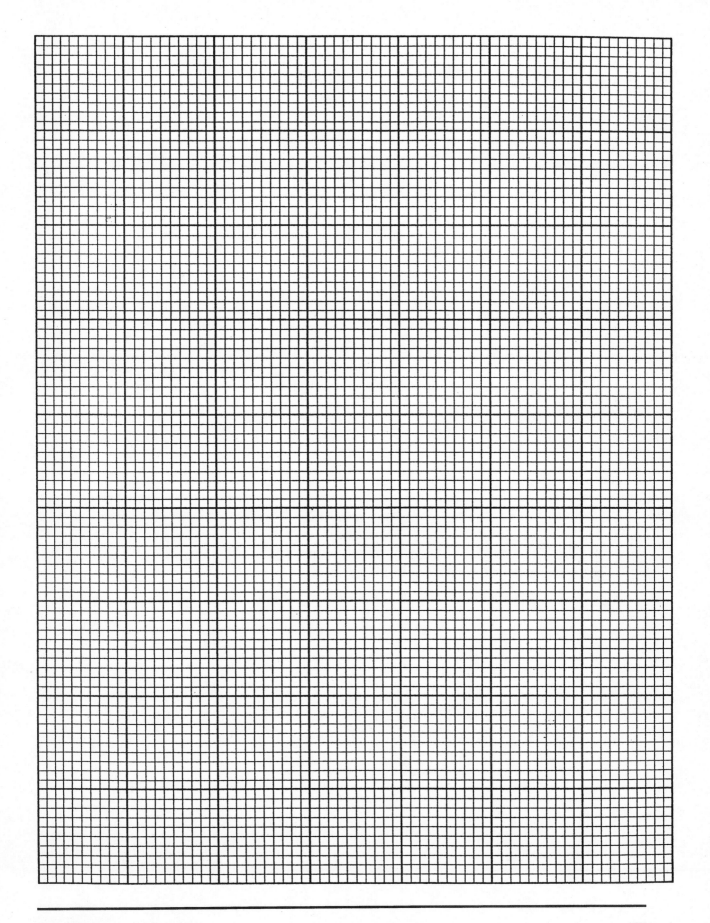

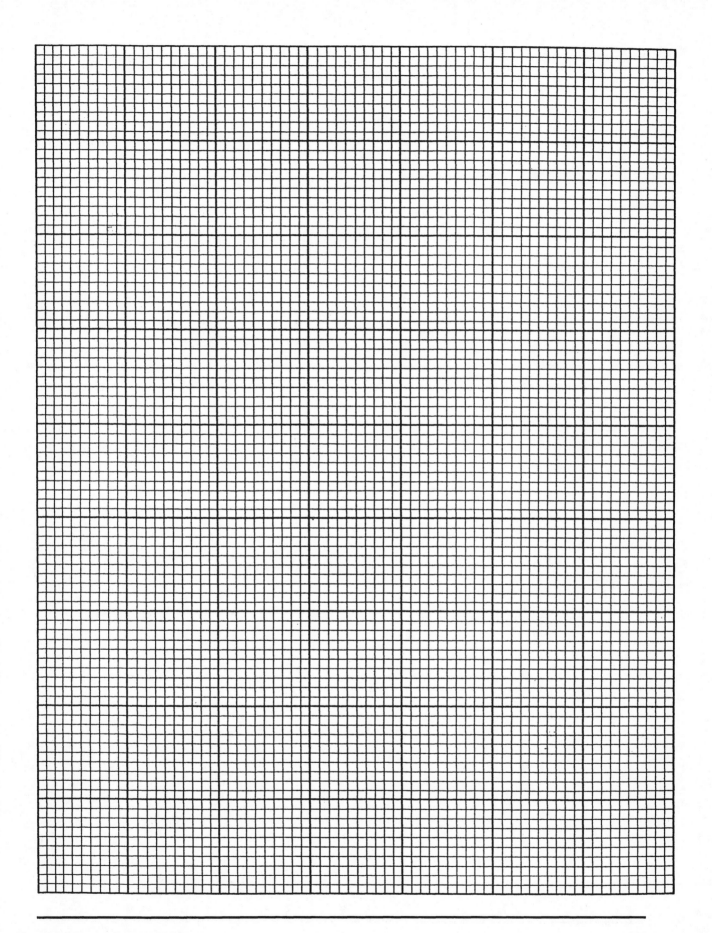

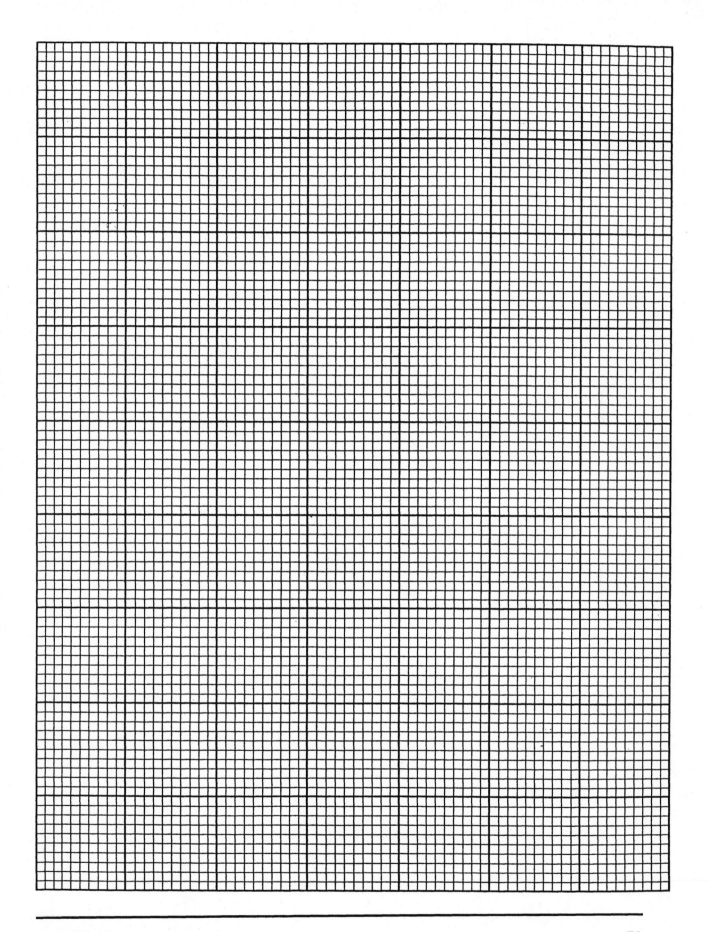

The Simple Pendulum

EXPERIMENT
10

INVESTIGATION OF DISTANCE, SPEED, AND ACCELERATION USING A SONIC RANGER

Name_____   Partner(s)_____

Date_____   _____

Section Number_____   _____

BASE CONCEPTS: Graphing

INTRODUCTION:

In physics, the concepts of distance, speed, and acceleration are important and closely related. _Speed_ (strictly speaking, average speed) is defined as the distance traveled by an object divided by the time of travel. _Acceleration_ (strictly, average acceleration) is defined as change in speed divided by the time for that change. This lab will allow you to investigate the meaning of these definitions and sharpen your understanding of them.

A sonar ranger measures the distance from it to any fairly large object in front of it at equal time intervals (typically every 1/60 second) by sending out sound pulses and determining the time it takes for their reflections to return. To get reliable data over a long period of time, the pulses must be reflected from something reasonably large and smooth. Rather than relying on your body to reflect the pulses, you will carry a sheet of poster board in front of you when doing the lab activities.

➤A.   In your own words, explain the meaning of the following terms:

▸ _Distance_ between two objects:

▸ *Speed* of an object

▸ *Acceleration* of an object

SUPPLIES: Computer, sonar ranger ("distance probe"), large poster board sheet.

PROCEDURE AND QUESTIONS:

1. First, investigate how the system works. Set the computer up so that distance as a function of time will be plotted on your computer screen.

2. Start about 0.5 m away from the sonar ranger and walk slowly and steadily away from it. Sketch the graph made by the computer. Be sure you pay attention to the units on the axes.

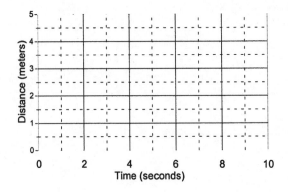

**Investigation of Distance, Speed
and Acceleration Using a Sonic Ranger**

➤B.	How would you expect the graph to differ from the one you just drew if you walked away from the distance probe more quickly? Why?

3.	Try it. Walk away from the 0.5 m line, again steadily, but more quickly than in Procedure 2.

➤C.	What does this graph look like in comparison with the one in Procedure 2?

➤D.	If there is disagreement between your prediction and actual results from Procedure 3, explain why you think the disagreement exists.

➤E. Now, predict the expected appearance of a graph of your starting 4 m in front of the sonar ranger and walking slowly toward the ranger. State *why* you made this prediction.

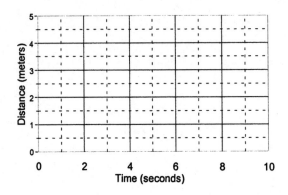

➤F. On the same graph, with a different type of line, predict what you expect walking rapidly toward the sonar ranger would look like and explain why.

4. Try Procedures E and F, starting at 4 m from the ranger. Make sketches of the computer graphs and, if there is any disagreement between the predictions and actual graphs, explain why.

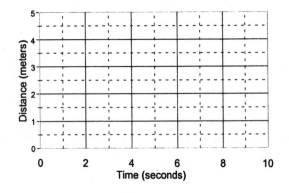
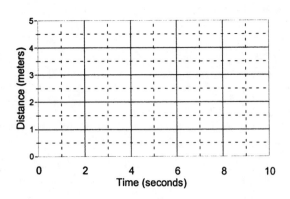

➢G. What do you look at on the computerized graph to determine if the person whose motion is being measured is moving toward or away from the sonar ranger?

➢H. What are the differences between the graphs for walking slowly and walking quickly?

➢I. How does the slope (steepness) of the line relate to your speed?

➢J. What does the type ("uphill" or "downhill") of slope signify?

Now that you have seen how the graphs are made, predict the following motion by drawing your prediction with a dotted line.

➤K. A person starts at the 0.5 m mark and stands there for 2 seconds, then slowly walks away from the sonar ranger to the 3-m mark, stands there for 2 seconds, then rapidly walks towards the sonar ranger back to the 1-m mark.

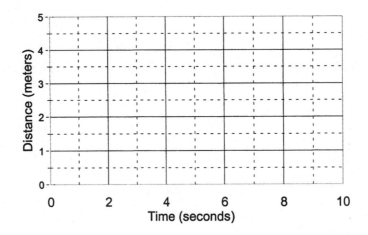

➤L. In groups of three and four, compare your predictions. Come up with **one** prediction that best represents your group's prediction.

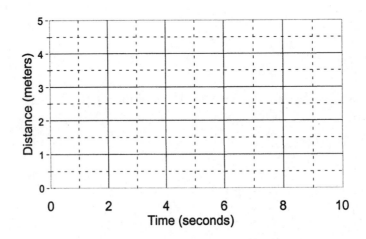

Investigation of Distance, Speed and Acceleration Using a Sonic Ranger

5. Now try it. Walk as described in Procedure K and make a sketch of the computer graph. If it does not agree with your group's prediction, explain why not.

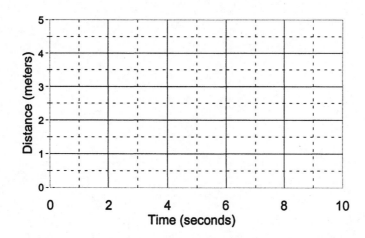

6. You are now ready to make graphs of varying shapes. There are three different graphs in the following section that you are to try to mimic. First, predict how you need to move to make each of the three and explain why you think these predictions are correct. **After** predicting all three motions, attempt to move in a way that will produce each graph on the computer. Write down how you moved and how your actual movements differed from those you predicted.

PREDICTION:

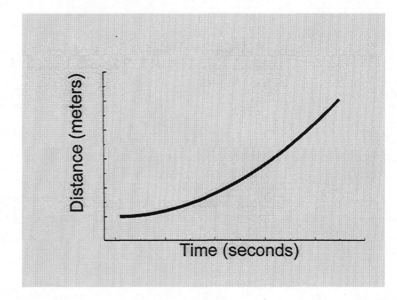

ACTUAL MOVEMENTS:

PREDICTION:

ACTUAL MOVEMENT:

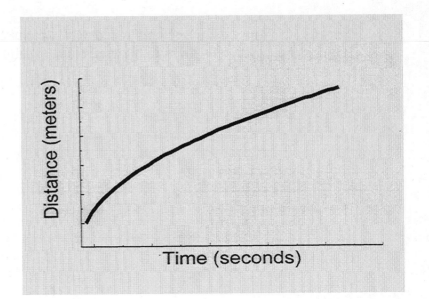

PREDICTION:

ACTUAL MOVEMENT:

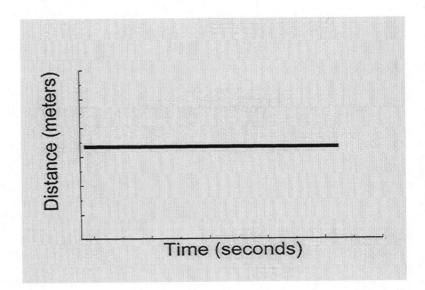

Investigation of Distance, Speed and Acceleration Using a Sonic Ranger

EXPERIMENT 11

A BOWLING BALL MEETS NEWTON'S SECOND LAW

Name_____   Partner(s)_____

Date_____   _____

Section Number_____   _____

BASE CONCEPTS: Force, Mass, Acceleration, Graphing Techniques

INTRODUCTION:

In this laboratory, you will explore the motion of a bowling ball when you hit it with evenly spaced light taps. The frequent light taps approximate a constant external force acting on a mass (the bowling ball). This approach can, therefore, be used to verify Newton's second law of motion, which states that the net force (F) acting on an object equals the mass (m) of the object times the object's acceleration (a): $F = ma$. (The acceleration is the change in velocity divided by time during which the change takes place.)

SUPPLIES:

Bowling ball, baseball, bat, several small sandbags (or other markers that don't bounce when dropped), 2-meter stick or steel tape. Each group also needs an unobstructed stretch of fairly level floor about 10 m long.

PROCEDURE:

1. Mark the point on the floor where you plan to start tapping the bowling ball. This point you call $X = 0$ and time $= 0$. You will drop sandbags to mark the location of the ball at equal time intervals. The time unit you will use in this lab is "number of taps." If you drop a sandbag every four taps, the total number of taps can easily be found by multiplying the number of dropped sandbags by four.

2. At the point $X = 0$, the person chosen to tap the ball should begin tapping. Tapping the ball lightly and regularly about three or four times a second works well. Every fourth tap, the person designated to drop sandbags should drop a sandbag on the floor next to the location of the ball at that tap (i.e., at the fourth tap, the eighth tap, etc.). The person tapping the ball should **be careful not to hit the ball harder as it begins to go faster**. You may want to practice for a few minutes before beginning to record data.

---

3.   After you run out of room (or sandbags), enter your data in a spreadsheet or in Table 1. The first column is the time in "number of taps" from the beginning (e.g., if it is the time for the third sandbag, the time in taps is given by:

$$\text{TIME} = 3 \text{ SANDBAGS} \times \frac{4 \text{ TAPS}}{\text{SANDBAG}} = 12 \text{ TAPS}).$$

In the second column you should enter the distance from $X = 0$ (where the ball started) to the sandbag corresponding to the time entered in that row of the table. The third column is the distance between consecutive sandbags. Some people find it easier to fill in the third column and then add the entries in the third column to generate the second column. See Table 2 on p. 91 for an example.

Table 1
Group Data

TIME (number of taps)	DISTANCE FROM POINT $X = 0$ (m)	DISTANCE FROM PREVIOUS SANDBAG (m)

Table 2
Sample Data

TIME (number of taps)	DISTANCE FROM POINT $X = 0$ (m)	DISTANCE FROM PREVIOUS SANDBAG (m)
0	0	--
4	5	5
8	12	7
12	20	8

4. You can graph your data in two ways. The second column corresponds to the total distance the ball has moved in the time shown in the first column. The third column corresponds to the change in distance between two adjacent sandbags at the time shown in the first column. If a constant force is applied to an object, according to Newton's second law a graph of velocity (change in distance divided by time for that change, which can be obtained by dividing the third column by 4 taps, giving units of meters/tap) versus elapsed time (in units of taps) should be a straight line. The slope of this line is equal to the acceleration. Alternatively, if the acceleration is zero (which means the net force on the object is zero), then by graphing the total distance (second column) versus elapsed time (first column), you should obtain a straight line with a slope equal to the velocity of the ball.

5. Either in a spreadsheet or by hand make two graphs: (1) total distance versus elapsed time and (2) speed (Column 3 divided by "4 taps") versus elapsed time.

GLOBAL QUESTIONS:

A. Did the speed of the ball decrease, increase, or remain the same as time increased?

How do you know?

B. Suppose you repeated the experiment striking the ball with more force at each tap. For a given number of taps, would the ball in this hypothetical case move faster, slower, or at the same speed as in the actual experiment?

Why?

C. Suppose you reran the actual experiment, this time leaving everything the same except using a less massive ball. Would the ball in this hypothetical case move faster, slower, or at the same speed as in the actual experiment?

Why?

D.   Which of your two graphs looks more like a representation of a straight line? Draw the best straight line you can (i.e., the line that passes closest to the data points) on this graph.

E.   Assuming that your short taps were equivalent to a constant force, does your observation in Question D verify Newton's second law ($F = ma = m \dfrac{\text{CHANGE IN SPEED}}{\text{CHANGE IN TIME}}$)?

Justify your answer.

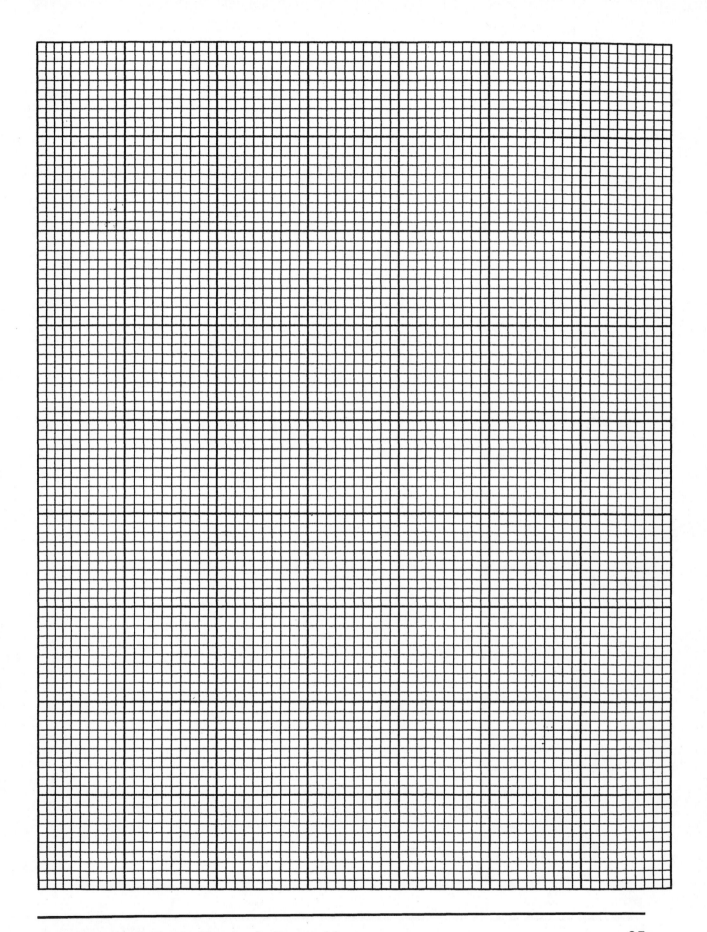

A Bowling Ball Meets Newton's Second Law

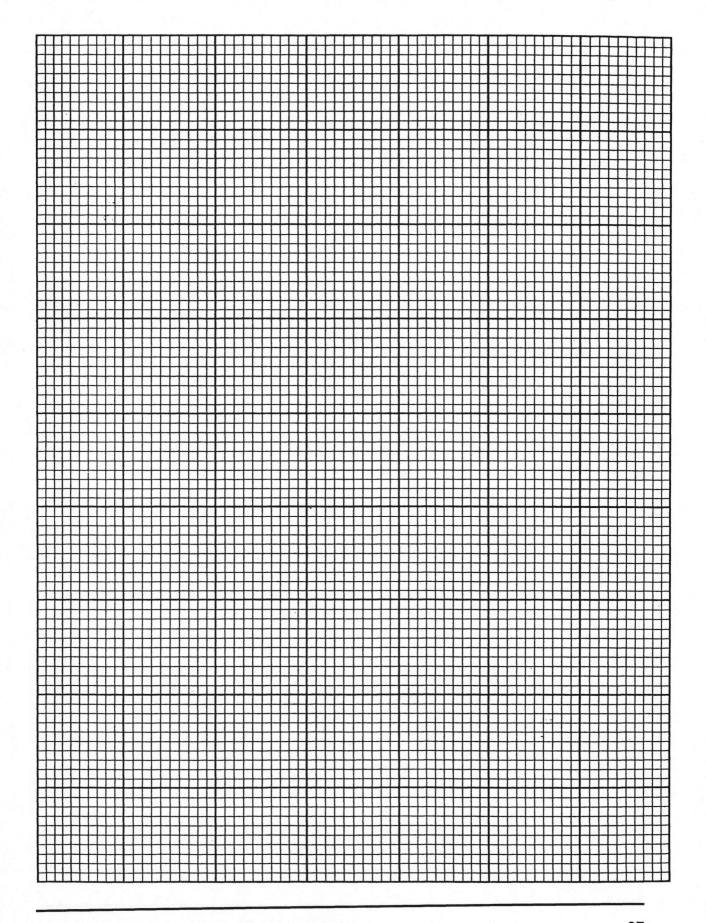

A Bowling Ball Meets Newton's Second Law

EXPERIMENT 12

NEWTON'S THIRD LAW: THE LAW OF FORCE PAIRS

Name_____   Partner(s)_____

Date_____   _____

Section Number_____   _____

BASE CONCEPTS: Force, Mass

INTRODUCTION:

Newton's third law of motion states that for every force exerted by an object A on an object B, object B exerts an equal force in the opposite direction on object A. *There are no known exceptions to this law* (though the meaning of "object" must be appropriately generalized in the context of modern physics). This law has many important ramifications, some of which will be explored here.

Following are some questions that relate to this law for you to answer. Answer them from your prior knowledge and experiences.

INITIAL QUESTIONS:

A.   If you leaned on the sink while standing on a bathroom scale, would the scale register a weight higher than, the same as, or lower than your weight? Why?

B.   If you try to hit a baseball with a bat, why is the completed swing less energetic if you hit the ball than if you do not hit it?

C. If you slap the wall hard, does it hurt your hand? Why?

D. Can you lift yourself up by grabbing onto the pants you are wearing and pulling? Why or why not?

You will investigate situations and problems like these in this lab.

SUPPLIES:

Tape, plastic soda straws, fishing line, large balloons, supports.

PROCEDURE AND QUESTIONS:

1. Thread a length of fishing line (about 30 cm longer than the distance between your two lab table supports) through two soda straw sections that are about 10–15 cm long. Tie the fishing line taut between the two supports.

2. Blow up one of the balloons, and while holding the balloon, let the air escape onto your arm.

➤E. Does the escaping air exert a force on your arm?

How do you know?

➤F. Do you think that the escaping air exerts a force on the air in the room? On the balloon? Why or why not?

3.   Let go of a full balloon while the air is escaping.

➤G.   Does the escaping air in the balloon exert a force on the room air? On the balloon?

       How do you know?

4.   Now blow the balloon up, and while one person in your group holds the air in the balloon, have another person tape the balloon to one of the drinking straw sections on the fishing line.

➤H.   What do you expect will happen when the balloon is freed and the air escapes? Why?

5.   Try it and see.

➤I.   Were your predictions correct?

       If not, what did happen?

6.   Now obtain a piece of cardboard about 3 in. square. Tape the cardboard to the other straw so that the balloon exhaust will hit it broadside.

➤J.   When the air is free to escape from the balloon, what motion (if any) do you predict for the straw to which the balloon is attached?

For the straw to which the cardboard is attached?

Why do you predict these outcomes?

7.   Try it.

➤K.   What does happen?

➤L.   Were your predictions correct? Why or why not?

8.   Next, blow up a balloon and tape it to the extreme left end of one straw so that the exhaust is pointing to the right. Tape the cardboard to the **same** straw so that the balloon's exhaust will again hit it broadside.

➤M.   When the air escapes from the balloon, what motion (if any) do you predict from the straw to which the balloon and cardboard are attached? Why?

9.   Try it.

➤N.   What did happen?

➤O.   Were your predictions correct? Why or why not?

GLOBAL QUESTIONS:

P.   As a group, list three examples of situations from your experience that are similar to the balloon and cardboard on different straws.   Be sure to explain *why* they are similar.

Q.   List three situations that are similar to the balloon and cardboard attached to the same straw. Explain *why* they are similar.

EXPERIMENT 13
STATIC FRICTION

Name_____   Partner(s)_____

Date_____   _____

Section Number_____   _____

BASE CONCEPTS: Force

INTRODUCTION:

A crucial step in the development of our fundamental understanding of mechanical phenomena was the recognition that friction is *not intrinsic* to motion: rather, it is one of many external forces that may act on a body and thereby influence its motion in accord with Newton's laws. This laboratory will provide you an opportunity to explore the behavior of one type of frictional force.

When the surfaces of two objects are in contact, the molecules at the surfaces can attract one another. Also, microscopic jagged edges of one surface can settle into the microscopic jagged edges of the other surface. Both of these effects contribute to the frictional force between the objects.

If one object is not moving relative to the other, the force that opposes any sliding motion is called the *static frictional force*. This force takes on whatever value is needed to compensate for any forces tending to cause the objects to slide past one another; however, as the slide-inducing forces increase, the static frictional force can increase only to a certain maximum value. If the slide-inducing forces increase further, the objects will begin to move past each other. In this lab you will investigate the dependence of the maximum static frictional force on various parameters.

SUPPLIES:

Two rectangular wooden blocks (one with an eye hook at one end), spring scale, meterstick, string.

PROCEDURE:

1. Your block has two sets of parallel sides that do not include the eye hook. In this lab, the larger of these is called "A" and the smaller "B" (see Figure 1 on the next page).

2. Measure the length and the width of both side A and side B and find their areas. (The area of a rectangle is length times width.)

Length of side A = _____.

Width of side A = _____.

Area of side A = _____.

Length of side B = _____.

Width of side B = _____.

Area of side B = _____.

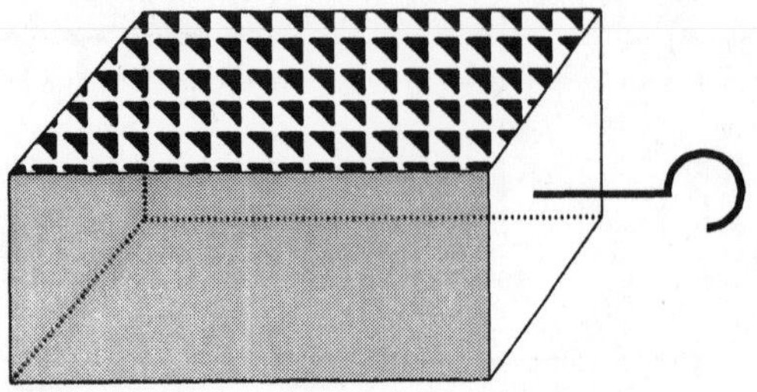

Side A -- larger of sides without hook

Side B -- smaller of sides without hook

Figure 1

3. Determine the weight of your block of wood in Newtons (N) by hanging it vertically from the spring balance.

The weight of the block is _____ N.

4. Place the block with one of the larger A sides on the lab table (see Figure 1).

5. **Connect the spring balance to the eye hook attached to the block. Keeping the spring balance parallel to the table**, pull the spring balance. Increase your pulling force until the block just begins to move. At this point the force you are exerting exactly balances the maximum static frictional force, so the two are equal. In the data table on p. 107 labeled "Lab Table and Side A," record the force you found that caused the block to just begin to move.

6. Repeat Procedure 5 four more times and record your results in the data table. Calculate the average of your five experimental values and enter it on the last line of the data table.

7. Repeat Procedures 5 and 6, this time placing smaller side B on the lab table. Enter your data in the data table "Lab Table and Side B" on p. 107.

Static Frictional Force

DATA TABLE Lab Table and Side A	
TRIAL #	FORCE THAT CAUSES BLOCK TO START
#1	
#2	
#3	
#4	
#5	
Average	

DATA TABLE Lab Table and Side B	
TRIAL #	FORCE THAT CAUSES BLOCK TO START
#1	
#2	
#3	
#4	
#5	
Average	

8. Now, secure a piece of sandpaper to the lab table and repeat Procedures 5 through 7 with the block placed on the sandpaper rather than directly on the lab table. Fill in the appropriate data tables. Remove the sandpaper when you are finished with this procedure and make sure that all the grit from the sandpaper is swept away from your experimental area.

DATA TABLE Sandpaper and Side A	
TRIAL #	FORCE THAT CAUSES BLOCK TO START
#1	
#2	
#3	
#4	
#5	
Average	

DATA TABLE Sandpaper and Side B	
TRIAL #	FORCE THAT CAUSES BLOCK TO START
#1	
#2	
#3	
#4	
#5	
Average	

9. Determine the weight of a second block.

The weight of the second block is _____ N.

10. Place the second block on top of the block you have been using. Repeat Procedures 5 through 7 with this combination and fill in the following data tables.

DATA TABLE Lab Table and Side A with 2 Blocks	
TRIAL #	FORCE THAT CAUSES BLOCKS TO START
#1	
#2	
#3	
#4	
#5	
Average	

DATA TABLE Lab Table and Side B with 2 Blocks	
TRIAL #	FORCE THAT CAUSES BLOCKS TO START
#1	
#2	
#3	
#4	
#5	
Average	

CALCULATIONS:

So that you can compare the various results you have obtained, copy information from your data tables to the following lines.

The weight of the first block = _____ N.

The weight of the first block + weight of the second block = _____ N.

The area of side A = _____ .

The area of side B = _____ .

The average force to start the block sliding with

Side A of one block on lab table = _____ .

Side B of one block on lab table = _____ .

Side A of one block on sandpaper = _____ .

Side B of one block on sandpaper = _____ .

Side A on lab table and two blocks = _____ .

Side B on lab table and two blocks = _____ .

Look at your values and answer the following questions. You can calculate the percent difference between two values much as you calculated percent error in other labs: find the positive difference between the two values, divide by the average of the two, then multiply final by 100%. Mathematically, this can be written as

$$\text{PERCENT DIFFERENCE} = \frac{|\text{VALUE \#1} - \text{VALUE \#2}|}{\frac{1}{2} \times (\text{VALUE \#1} + \text{VALUE \#2})} \times 100\%.$$

GLOBAL QUESTIONS:

A. What is the percent difference of the maximum static frictional forces when you use side A and side B on the lab table?

B. What is the percent difference of the maximum static frictional forces when you use side A and side B on the sandpaper?

C. What is the percent difference of the maximum static frictional forces when you use side A and side B on the lab table with the extra block on top?

D.   What is the percent difference of the maximum static frictional forces when you use side A on the lab table and side A on the sandpaper?

E.   What is the percent difference of the maximum static frictional forces when you use side B on the lab table and side B on the sandpaper?

F.   In what way did the extra mass on top of the original block affect the results? If you had put a heavier second block on top, would you expect a larger or smaller maximum static frictional force? Why?

G.   For a given lab setup (i.e., sandpaper or no sandpaper, one block or two blocks), in what significant ways did the results depend on whether side A or side B was used?

EXPERIMENT 14

KINETIC FRICTION

Name_____   Partner(s)_____

Date_____   _____

Section Number_____   _____

BASE CONCEPTS: Force

INTRODUCTION:

Crucial to understanding mechanical phenomena at a fundamental level is the recognition that friction is *not intrinsic* to motion. Friction is simply treated as an external force that influences motion in accord with Newton's laws. This laboratory will provide you an opportunity to explore the behavior of on type of frictional force.

When the surfaces of two objects are in contact, the molecules at the surfaces can attract one another. Also, microscopic jagged edges of one surface can settle into the microscopic jagged edges of the other surface. Both of these effects contribute to the frictional force between the objects.

If one of the objects is moving relative to the other, the frictional force that opposes this sliding motion (i.e., it acts in a direction opposite the motion) is called the *kinetic frictional force*. In this lab you will investigate the dependence of the kinetic frictional force on various parameters.

SUPPLIES:

Two rectangular wooden blocks (one with an eye hook at one end), spring scale, meterstick, string.

PROCEDURE:

1. Your block has two sets of parallel sides that do not include the eye hook. In this lab, the larger of these is called "A" and the smaller "B" (see Figure 1 on the next page).

2. Measure the length and the width of both side A and side B and find their areas. (The area of a rectangle is length times width.)

Length of side A = _____.

Width of side A = _____.

Area of side A = _____.

Length of side B = _____.

Width of side B = _____.

◥◥◥	Side A -- larger of sides without hook
(shaded)	Side B -- smaller of sides without hook

Figure 1

Area of side B = _____.

3.   Determine the weight of your block of wood in newtons (N) by hanging it vertically from the spring balance.

The weight of the block is _____ N.

4.   Place the block with one of the larger A sides on the lab table (see Figure 1).

5.   **Connect the spring balance to the eye hook attached to the block. Keeping the spring balance parallel to the table,** pull the spring balance connected to the block. Increase your pulling force until the block begins to move. Once the block starts to move, find the force it takes to keep the block moving *at a constant speed*. Under these conditions the force you are exerting exactly balances the kinetic frictional force, so the two are equal. In the data table on p. 113 labeled "Lab Table and Side A," record the force you found that caused the block to move at a constant speed.

6.   Repeat Procedure 5 four more times and record your results in the data table. Calculate the average of your five experimental values and enter it on the last line of the data table.

7.   Repeat Procedures 5 and 6, this time placing smaller side B on the lab table. Enter your data in the data table "Lab Table and Side B" on p. 113.

DATA TABLE Lab Table and Side A	
TRIAL #	FORCE THAT CAUSES BLOCK TO SLIDE
#1	
#2	
#3	
#4	
#5	
Average	

DATA TABLE Lab Table and Side B	
TRIAL #	FORCE THAT CAUSES BLOCK TO SLIDE
#1	
#2	
#3	
#4	
#5	
Average	

8. Now, secure a piece of sandpaper to the lab table and repeat Procedures 5 through 7 with the block placed on the sandpaper rather than directly on the lab table. Fill in the appropriate data tables. Remove the sandpaper when you are finished with this procedure and make sure that all the grit from the sandpaper is swept away from your experimental area.

DATA TABLE Sandpaper and Side A	
TRIAL #	FORCE THAT CAUSES BLOCK TO SLIDE
#1	.
#2	
#3	
#4	
#5	
Average	

DATA TABLE Sandpaper and Side B	
TRIAL #	FORCE THAT CAUSES BLOCK TO SLIDE
#1	
#2	
#3	
#4	
#5	
Average	

9. Determine the weight of a second block.

The weight of the second block is _____ N.

10. Place the second block on top of the block you have been using. Repeat Procedures 5 through 7 with this combination and fill in the following data tables.

DATA TABLE Lab Table and Side A with 2 Blocks	
TRIAL #	FORCE THAT CAUSES BLOCKS TO SLIDE
#1	
#2	
#3	
#4	
#5	
Average	

DATA TABLE Lab Table and Side B with 2 Blocks	
TRIAL #	FORCE THAT CAUSES BLOCKS TO SLIDE
#1	
#2	
#3	
#4	
#5	
Average	

CALCULATIONS:

So that you can easily compare the various results you have obtained, copy information from your data tables to the following lines.

The weight of the first block = _____ N.

The weight of the first block + weight of the second block = _____ N.

The area of side A = _____ .

The area of side B = _____ .

The average force to keep the block moving at constant speed with

Side A of one block on lab table = _____ .

Side B of one block on lab table = _____ .

Side A of one block on sandpaper = _____ .

Side B of one block on sandpaper = _____ .

Side A on lab table and two blocks = _____ .

Side B on lab table and two blocks = _____ .

Look at your values and answer the following questions. You can calculate the percent difference between two values much as you calculated percent error in other labs: find the positive difference between the two values, divide by the average of the two, then multiply by 100%. Mathematically, this can be written as

$$\text{PERCENT DIFFERENCE} = \frac{|\text{VALUE \#1} - \text{VALUE \#2}|}{\frac{1}{2} \times (\text{VALUE \#1} + \text{VALUE \#2})} \times 100\%.$$

GLOBAL QUESTIONS:

A.   What is the percent difference of the average kinetic frictional forces when you use side A and side B on the lab table?

B.   What is the percent difference of the average kinetic frictional forces when you use side A and side B on the sandpaper?

C.   What is the percent difference of the average kinetic frictional forces when you use side A and side B on the lab table with the extra block on top?

D.   What is the percent difference of the average kinetic frictional forces when you use side A on the lab table and side A on the sandpaper?

E.   What is the percent difference of the average kinetic frictional force when you use side B on the lab table and side B on the sandpaper?

F.   In what way did the extra mass on top of the original block affect the results? If you had put a heavier second block on top, would you expect a larger or smaller kinetic frictional force? Why?

G.   For a given lab setup (i.e., sandpaper or no sandpaper, one block or two blocks), in what significant ways did the results depend on whether side A or side B was used?

H.   Repeat Procedures 5 and 6 to see if the constant speed that you chose makes a difference. Try a slow, a medium, and a fast speed with side A on the lab table. See if the average force to keep the block moving changes with speed. Summarize your results below (including your data).

(OPTIONAL) - Comparing Static and Kinetic Frictional Forces

Note: To complete this section you must have completed both this lab and the static friction lab. Ask your instructor if you are not sure.

I.   In the following chart circle which force was larger — the force necessary to start the block moving or the force needed to keep the block moving at a constant speed — for each of the following cases.

a) Side A -- lab table

b) Side B -- lab table

c) Side A -- sandpaper

d) Side B -- sandpaper

e) Side A -- two blocks

f) Side B -- two blocks

starting force	moving force
starting force	moving force
starting force	moving force
starting force	moving force
starting force	moving force
starting force	moving force

Is there a pattern in the table?

What accounts for this pattern? (*Hint:* See introductions to the two frictional force labs.)

J.   List three situations in which a frictional force is useful. In each case explain why it is useful.

K.   List three situations in which a frictional force is detrimental. In each case explain why it is detrimental.

EXPERIMENT 15

OBSERVATIONS OF AIR PRESSURE

Name_____   Partner(s)_____

Date_____   _____

Section Number_____   _____

BASE CONCEPTS: Force

INTRODUCTION:

Air pressure is the perpendicular force per unit area that air exerts on an object. In most everyday occurrences, you probably do not explicitly notice the effects of air pressure; however, there are many indications of its presence around you. For example, your lungs expand when you breathe in. When you blow into a balloon, it expands due to the increased air pressure inside.

What happens to an object if the pressure on one side of the object is greater than that on the other side? In general, if the air pressure on one side of a barrier is larger than the air pressure on the other side, there is a net force toward the smaller pressure.

➤A.   Using the given definition of pressure, describe what you think happens if the pressure on one side of an expandable object is larger than the pressure on the other side. Why?

This lab consists of a series of activities that let you "see" the effects of air pressure.

SUPPLIES:

Drinking glass with a smooth edge, index cards large enough to completely cover the opening of the glass, bowl or tray with sides at least 5 in. high, hot plate or Bunsen burner, empty soda can, bowl with about 5 in. of cold water in it, insulated glove.

➤B.   Can you put water in the glass and, using air pressure, keep the water in the glass when you turn the glass upside down? How would you try this with the materials listed for this lab?

Carefully try your idea and see if it works? Did it?

PROCEDURE AND QUESTIONS:

One way you can keep water in a glass when it is turned upside down using air pressure is to:

1.   Place the glass in the bowl and fill the glass with water until it overflows.

2.   Place your index card on top of the filled glass so that the card completely covers the opening of the glass.

3.   While pushing the index card against the top of the glass, turn the glass over and then remove your hand from the index card. Do not remove the index card. *Just in case, invert the glass over the bowl.*

➤C.   Why did the water stay in the glass?

➤D.   On which side of the index card is the air pressure greater?

How do you know?

➢E.   What causes the pressure on the outside?

4.   Now you are starting another activity that also demonstrates the presence of air pressure. Take the soda can and put a few teaspoons of water inside.

5.   Heat the can until steam vigorously comes out of the top.

6.   With the bowl placed close to where the heating occurs, and using a gloved hand, quickly place the heated can in the cold water *with its open end down*.

      What happened?

➢F.   Did the inside or the outside of the soda can have a larger air pressure?

      How do you know?

➢G.   What happened *during the heating process* that made the answer to the previous question come out as it did?

➤ H.   What happened when you put the can in the cold water to cause the can to do what it did?

GLOBAL QUESTIONS:

Using the given definitions and your experimental observations, explain on the following pages any three of the following five physical observations

1.   The gauge pressure (i.e., the difference between the pressure inside the tire and the pressure outside of it) for a given tire becomes larger as you move to the mountains from sea level, although no air is added to or removed from the tire.

2.   If soda is put in a thin plastic container made for milk and the lid is replaced, the container bulges on the sides.

3.   You are cooking soup in a pan on the stove. When the soup begins to boil, you put a lid on the pan and remove it from the heat. After the pan cools for a while, the lid is very difficult to remove.

4.   You need two holes in a can containing liquid for the contents to empty smoothly.

5.   Airplane runways in Denver, Colorado, (altitude about 5000 ft above sea level) are longer than in Boston, Massachusetts, (altitude about 100 ft above sea level). (Think about what enables an airplane to fly.)

I.

I. (continued)

J.

K.

EXPERIMENT 16 — STORING ENERGY

Name_____   Partner(s)_____

Date_____   _____

Section Number_____   _____

BASE CONCEPTS: Work, Energy, Forms of Energy

INTRODUCTION:

In order to transfer energy to an object, work must be done on that object. Here work has the very specific definition of external force exerted on the object multiplied by the distance the object moves in the direction of that the force is exerted.

➤A.   This definition guarantees that energy is transferred from the object doing the work to the object on which work is done. Why?

When positive work is done on an object, any of several things can happen, including the following.

1.   The object may, in turn, do work on another object; that is, it can give away the energy it gained to its surroundings. (An example of this would be hammering a nail into wood. The energy you transfer to the nail ends up as thermal energy in the wood and the nail.)
2.   The object's speed may increase: a gain in kinetic energy.
3.   The object's temperature may rise, indicating that energy has been stored internally: a gain in thermal energy.
4.   The object may store the energy for later use by changing its shape or configuration: a gain in elastic (potential) energy. (An example of this is the stretching of a spring when a door is opened. The energy used to stretch the spring is stored as elastic (potential) energy in the stretched spring. When the door is released, the spring returns to its original length and does work on the door, causing it to close.)
5.   The object may rise in the earth's gravitational field: a gain in gravitational (potential) energy.

SUPPLIES:

Some or all of the following: rubber balls of various sizes, balloons, windup toys, push forward cars, pullback cars, books, rubber bands, sling shot, pop-ups, pulley, mass holders, slotted masses, string.

Your lab may not include all of these, so your instructor will tell you if you should skip some parts or substitute alternative "toys."

PROCEDURE AND QUESTIONS:

Each of the following sections is quite similar, but there are subtle differences you should watch for as you proceed. You can tell if you have done work on a body in either of two ways. If you exerted a force on the object, and this caused the object to move in the direction of that force, **you did work**. If you did not exert a force, you could not have done any work. Alternatively, if there is energy in the body that was not there before you did something to it, **you did work**. In using this second approach, you must be careful to ensure that the object's energy actually increased. Don't be fooled by situations in which you exert no force on the object but just "trigger" a transformation of the object's energy from one form to another (e.g., from gravitational potential energy to kinetic energy) with no increase in the object's total energy.

There are no detailed procedures to be followed for this lab. Work in groups of two or three to explore how energy is stored by each of the suggested physical systems (or objects). Discuss each set of observations, ideas, and answers until you reach a group consensus. Also, **please be careful**; stored energy should always be handled with caution.

1.   Explore energy storage in a balloon.

➤B.   Do you do work when you blow up a balloon? How do you know?

➤C.   How is energy stored in a balloon? In what form?

➣D.   How can you get the energy back from the balloon?

➣E.   How could the balloon do work on a different object by using its stored energy?

2.   Explore energy storage in a windup toy.

➣F.   Do you do work when you wind up the toy?   How do you know?

➣G.   How is energy stored in the windup toy? In what form?

➣H.   How can you get the energy back from the windup toy?

➤I.   How could the windup toy do work on a different object by using its stored energy?

3.   Explore energy storage in a rubber band.

➤J.   Do you do work when you stretch a rubber band?   How do you know?

➤K.   How is energy stored in a rubber band? In what form?

➤L.   How can you get the energy back from the rubber band?

➤M.   How could the rubber band do work on a different object by using its stored energy? (Please do not try your hypothesis on this one!)

4. Explore energy storage in a book.

➤N. Do you do work when you raise a book from the floor to the lab table? How do you know?

➤O. How is energy stored in the book? In what form?

➤P. How can you get the energy back from the book?

➤Q. How could the book do work on a different object by using its stored energy?

5. Explore energy storage in a popup. What is a popup? A popup can be made by taking a racquetball, cutting it in half, and carefully cutting away the sides. If correctly made, an inverted popup will remain inverted until it is dropped. When dropped, it strikes the floor, reverts to its original shape, and pops up. If you fail to cut away enough of the sides, it will not revert when dropped. If you cut away too much of the sides, it will revert before striking the floor.

➤R. Do you do work when you invert the popup? How do you know?

➤S. How is energy stored in a popup? In what form?

➤T. How can you get the energy back from the popup?

➤U. How could the popup do work on a different object by using its stored energy?

6. Explore energy storage in an Atwood machine. What is an Atwood machine? An Atwood machine consists of two unequal masses connected by a string over a pulley (see Figure 1). When released, the larger mass falls as the smaller mass rises.

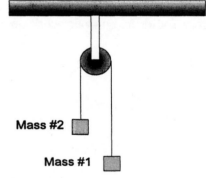

Mass #2

Mass #1

Figure 1

➤V. Do you do work when you let the string go so that the larger mass can fall? How do you know?

➤W.   How is energy stored in an Atwood machine? In what form?

➤X.   How can you get the energy back from the Atwood machine?

➤Y.   How could the Atwood machine do work on a different object by using its stored energy?

7.   Explore energy storage in pullback toys. These are toys (usually cars) made so that when you put them down and pull them back and let go, they move forward.

➤Z.   Do you do work when you pull back this toy?   How do you know?

➤AA. How is energy stored in the toy? In what form?

➢BB. How can you get the energy back from the toy?

➢CC. How could the toy do work on a different object by using its stored energy?

8.   Explore energy storage in push-forward cars. These are toy cars made so that when you push them forward on the floor a few times and then let them go, they move forward.

➢DD. Do you do work when you push the toy car forward?   How do you know?

➢EE. How is energy stored in this toy? In what form?

➢FF. How can you get the energy back from this toy?

➤GG. How could the toy do work on a different object by using its stored energy?

9.   Explore energy storage by a ball.

➤HH. Do you do work when you drop a ball? How do you know?

➤II.   How is energy stored in a ball when it hits the ground? In what form?

➤JJ.   How can you get the energy back from the ball?

➤KK. How could the ball do work on a different object by using its stored energy?

GLOBAL QUESTION:

LL.   What are two other ways that you could either store energy or use it by doing work on some object around you?

EXPERIMENT 17

WORK, ENERGY TRANSFER AND POWER

Name_____   Partner(s)_____

Date_____   _____

Section Number_____   _____

BASE CONCEPTS: Forms of Energy

INTRODUCTION:

In physics the term *work* has a very specific definition. According to this definition, you do work on an object if and only if the object moves in the direction in which you exert a force on it. By this definition, Atlas did no work on the world when he held it up! If you do exert a force on an object and the object moves in the direction of the applied force, you transfer energy from you to the object, and the work you do equals the energy you transfer to the object.

➤A.   In your own words, explain the meaning of *work* (in physics). Give an example if it helps.

➤B.   You step into an elevator and stand there on the elevator floor, deciding which building level you want. Is the elevator floor doing work on you? Why or why not?

➤C. You decide to go up to the observation level and push the button. Once the elevator starts moving up, is it doing work on you? Why or why not?

Notice that the physics definition of work does not involve time. If you lift a book a specific distance, you impart the same amount of energy to it regardless of the time it takes to lift the book that distance the book gains gravitational potential energy equal to its weight times the vertical distance you lifted it; however, your *rate* of energy output or *power* output (amount of energy transferred [i.e., work done] divided by the time) does differ. The shorter the time taken to transfer a given amount of energy, the larger the power output (or input). For example, a car with a higher horsepower (a unit of power) engine can accelerate more quickly than a similar one with a lower horsepower engine. The larger engine can do more work in a given amount of time. Power has units of watts (1 W = 1 J/s) or horsepower (1 hp = 746 W = 550 lb-ft/s).

A person's body can do work on itself. Your legs can lift your entire body up a flight of stairs. Your arms can push your body off the floor when doing push-ups. The power that a body can put out depends on many things, including the duration of the task. No long distance runner runs as fast as a sprinter. A sprinter produces a higher power output, but only for a short period of time. In this lab you will measure your energy and power output in climbing a set of stairs.

SUPPLIES: Set of stairs, meterstick, timer.

PROCEDURE:

1. Circle the range on the left in which your weight falls and use the weight to the right of that range for your calculations.

   under 100 lb   use 90 lb = 400 N

   100-150 lb   use 125 lb = 556 N

   150-200 lb   use 175 lb = 778 N

   over 200 lb   use 225 lb = 1001 N

2.   Measure the vertical height of the set of stairs you will use for this experiment. In this lab you are primarily interested in approximate results, so carefully measuring the vertical height of one step and multiplying by the number of steps will give you a reasonable estimate.

The vertical height of one step is _____ m.

The number of steps is _____.

The total height of the staircase is _____ m = _____ ft.
       (1 m = 3.28 ft)

3.   If you climb this set of stairs, the amount of energy transferred to your body by your legs equals your weight times the vertical distance moved.

The work done = energy transferred is _____ N-m = _____ lb-ft.

4.   You should all work in groups of two or three. You are to slowly walk up the stairs while your partner(s) measure the time it takes for you to do so.

The time it took me to walk up the stairs was _____ seconds.

The time it took partner #1 to walk up the stairs was _____ seconds.

(The time it took partner #2 to walk up the stairs was _____ seconds.)

5.   From the information in Procedures 3 and 4, calculate the power output for each person in your group. (1 W = 1 J/s = 1 N-m/s, and 1 hp = 550 lb-ft/s.)

The power output for my walking up the stairs was _____ W = _____ hp.

The power output for partner #1 walking up was _____ W = _____ hp.

(The power output for partner #2 walking up was _____ W = _____ hp.)

6.   Now, each person should repeat Procedures 4 and 5, this time running as fast as **safely** possible rather than walking.

The time it took me to run up the stairs was _____ seconds.

The time it took partner #1 to run up the stairs was _____ seconds.

(The time it took partner #2 to run up the stairs was _____ seconds.)

The power output for my running up the stairs was _____ W = _____ hp.

The power output for partner #1 running up was _____ W = _____ hp.

(The power output for partner #2 running up was _____ W = _____ hp.)

GLOBAL QUESTIONS:

D.   Would your power output **when walking** have changed much if you had walked up three or four flights of stairs rather than one? Why or why not?

E.   Would your energy output **when walking** have changed much if you had walked up three or four flights of stairs rather than one? Why or why not?

F.   Would your power output **when running** have changed much if you had run up three or four flights of stairs rather than one? Why or why not?

Work, Energy Transfer and Power

EXPERIMENT 18   TEMPERATURE AND HEATING

Name_____   Partner(s)_____

Date_____   _____

Section Number_____   _____

BASE CONCEPTS: Temperature, Energy

INTRODUCTION:

Suppose you have a large mug of hot chocolate (or coffee, if you prefer), which initially has a temperature of 90°C. You leave it sitting in a room within a temperature of 23°C.

➤A.   Will the temperature of the hot chocolate increase or decrease as it sits in the room? How do you know? Does your answer depend at all on the temperature of the room?

In the example, how hot or cold will the chocolate become?

➤B.   Does the temperature of the hot chocolate change *more rapidly* when it is first placed in the room or after it has been in the room for a while? Why?

Suppose you have a large glass of cold lemonade, initially at 5°C, in the same room.

➤C.   Will the temperature of the cold lemonade increase or decrease as it sits in the room?

How hot or cold will the lemonade become?

➤D. Does the temperature of the cold lemonade change *more rapidly* when it is first placed in the room or after it has been in the room for a while? Why?

From these questions, you can see that something must be transferred to or from the room in order for the temperatures of the beverages to change. That something is commonly called *heat*, and its spontaneous flow always takes place from an object at a higher temperature to an object at a lower temperature. Heat is thermal energy transferred from one object to another because of a temperature difference.

In this lab you will investigate heat transfer from hot water to the surrounding room. You will be using hot water, so **please be careful** not to burn yourself (or anyone else, for that matter).

SUPPLIES:

Computer, computerized temperature probes, and cup of hot water.

PROCEDURE AND QUESTIONS:

1. Discuss your prediction question answers (given earlier in this lab) as a group and make sure that your group reaches a consensus concerning the answers.

2. Decide as a group which graph in Figure 1 best represents the way the temperature of a hot liquid sitting in a cool room will change as a function of time.

➤E. Justify your answer by explaining **why** you picked the graph you did.

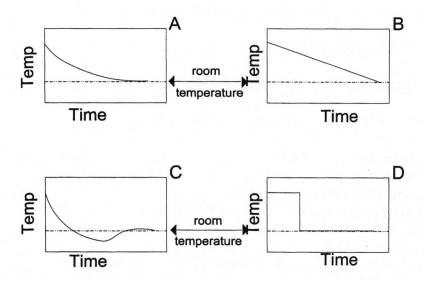

Figure 1

3.   Now, do the experiment. Set your computer thermometer temperature axis for a 20°C to 85°C range. Set your time axis duration for 20 minutes.

4.   Calibrate the thermometer if necessary.

5.   Position your thermometer in a cup of hot water (do not use a Styrofoam cup) so that it measures the water temperature, and start the computer reading the temperature.

6.   After your computer has collected the data, look at the graph on the computer screen. If necessary, change the range of the temperature axis so you can see the trend better.

7.   On the graph supplied in Figure 2 carefully sketch the graph of temperature versus time (or use the computer to print a copy of your graph).

8.	Obtain the following information from the graph.

The water temperature when the readings started was _____ °C.

The water temperature 5 minutes after the start was _____ °C.

The water temperature 15 minutes after the start was _____ °C.

The water temperature 20 minutes after the start was _____ °C.

9.	Calculate the difference in temperature during the first 5 minutes: _____ °C.

Between 5 and 10 minutes: _____ °C.

Between 10 and 15 minutes: _____ °C.

Between 15 and 20 minutes: _____ °C.

➤F.	Did the temperature change more in the first 5 minutes or in the last 5 minutes?

Does this agree with your prediction?   Why or why not?

➤G.	Which of the four graphs in Figure 1 looks most like your data? (Make sure that the temperature scale on your computer is sensitive enough to see which one it looks like. If necessary, go to a more sensitive temperature scale.)

Does this agree with your prediction?   Why or why not?

Temperature and Heating

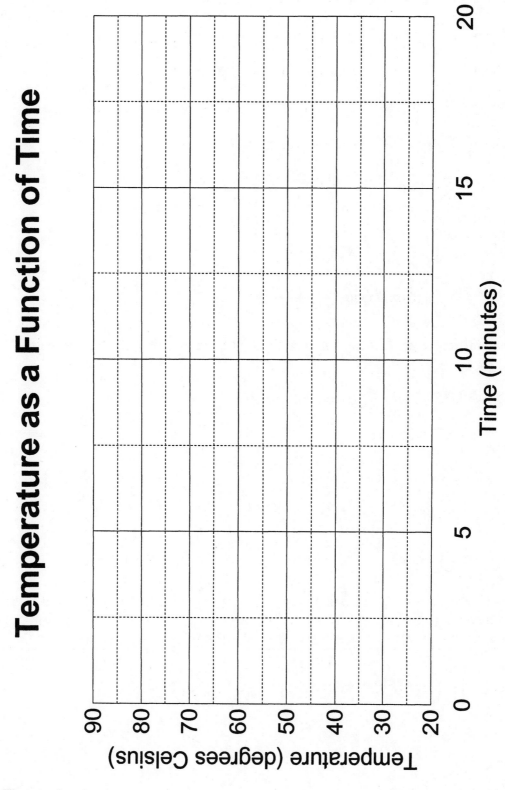

Figure 2

Temperature and Heating

EXPERIMENT 19

SPECIFIC HEAT CAPACITIES OF VARIOUS MATERIALS

Name_____   Partner(s)_____

Date_____   _____

Section Number_____   _____

BASE CONCEPTS: Graphing, Heat, Thermal Energy, Calculations

INTRODUCTION:

After answering these questions, it should become clear to you that temperature and the sensation of "hot" or "cold" are not necessarily the same. An object *feels* hot if, when you touch it, it transfers heat to you. It *feels* cold if you transfer heat to it. If very little heat transfers between you and the object, then you sense that the object has about the same temperature as you.

➤A.   What are the temperatures of two different nonliving objects after they sit in a 70°F room for 24 hours?

Do the final temperatures of the objects depend on the materials of which they are made? How do you know?

➤B.   When you get up in the morning, you step barefoot into your bathroom. You place your right foot on a shag rug and your left foot on the tile floor. Which material *feels* colder — the rug or the tile floor?

Which *is* colder — the rug or the tile floor? (Refer to your answer in Question A for help here.)

➤C.   Can anything hotter than your hand *feel* cold to your hand? Justify your answer.

➤D.   Returning to Question B, and referring to the initial paragraph, does your foot transfer more thermal energy to the rug or to the tile? How do you know?

In the preceding questions the amount of heat transferred depends on the temperature of the material on which you stand, how readily the material conducts heat (its thermal conductivity), and the material's ability to store thermal energy (its specific heat capacity). In this lab you will investigate the specific heat capacities ("specific heats," for short) of various materials. The magnitude of the specific heat of a material indicates how much thermal energy a specific mass of the material can store for a given temperature change. Since heat is a flow of thermal energy, the larger the specific heat of a material, the more heat can be transferred to or from the material for a given temperature change. Mathematically, the specific heat capacity C of a material is given by

$$C = \frac{Q_{in}}{m \times \Delta T},$$

where Q_{in} is the amount of heat transferred into an object made of the material (energy units),

m is the mass of the object (mass units),

ΔT is the temperature change of the object due to the heat transfer (temperature units).

Something with a large heat capacity is able to absorb or give off a lot of heat with very little change in temperature. A baking potato has a large specific heat capacity. It takes a long time to heat it up, and it also takes quite a while for the potato to cool.

➤E.　What common material has a large specific heat capacity?

➤F.　What common material has a small specific heat capacity?

➤G.　Does the rug or the tile in Question B have a higher specific heat capacity? Justify your answer.

To measure the specific heat capacities of materials, the *principle of thermal energy transfer* (sometimes called the *zeroth law of thermodynamics*) and energy conservation (or the *first law of thermodynamics*) will be used. In this lab you will create a system consisting of your material sample and hot water.

The water, being hotter than the material you are studying, gives up an amount of heat equal to the mass of the water multiplied by its specific heat times its temperature change. This may be written in equation form as

$$Q_{out} = m_w \, C_w \, (\text{Initial } T_w \; - \; \text{Final } T_w),$$

where the *w* stands for water, and *T* is temperature.

The heat absorbed by the material ("mat") you are investigating is given by

$$Q_{in} = m_{mat} C_{mat} (\text{Final } T_{mat} \; - \; \text{Initial } T_{mat}).$$

If the system is adequately insulated, energy conservation requires that the heat lost by the water is equal to the heat absorbed by the sample: $Q_{in} = Q_{out}$. For water $C_w = \dfrac{1 \text{ cal}}{\text{gram} - C°}$. Thus, by measuring the masses and temperatures, you can calculate the specific heat capacity of the material under investigation.

SUPPLIES:

Various metal specimens at room temperature, Styrofoam cups, hot water, thermometers (preferably computerized).

Specific Heats of Various Materials　　　　　　　　　　　　　　　　　　**149**

PROCEDURE AND QUESTIONS:

1. Using a thermometer, find the temperature of the room.

    The room temperature is _____ °C.

2. Choose two different specimens to investigate. Do not handle them any more than necessary to ensure that they stay at room temperature until you drop them in the water.

3. Measure and record the mass of your Styrofoam cup.

    The mass of the Styrofoam cup is _____ grams (g).

4. Measure and record the mass of the first specimen.

    The type of material of specimen #1 is _____.

    The mass of specimen #1 is _____ g.

5. Place about 100 ml (roughly ½ cup) of hot water (from a coffeemaker, microwave, or other appropriate source) in the Styrofoam cup.

6. Determine the mass of the water by measuring the mass of the cup with the water and subtracting the mass of the empty cup.

    The mass of the cup and water is _____ g.

    The mass of the water is _____ g.

7. On the computer set the temperature range to 25°C–90°C. Set the measuring time duration to 7 minutes (420 seconds).

8. Place the thermometer in the water and begin collecting temperature data. After 2 or 3 minutes, carefully place specimen #1 in the water. Make sure no water splashes out as you do this.

9. Let the computer continue to take temperature measurements for another 4 minutes.

Specific Heats of Various Materials

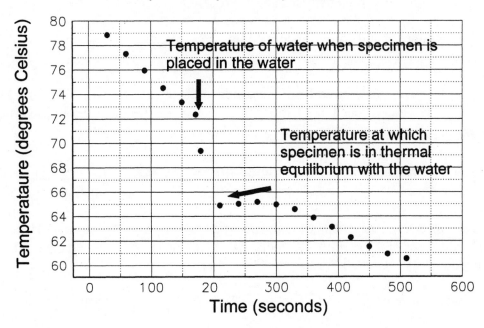

Figure 1

10. Print the graph of temperature versus time. The graph is Figure 1 gives an example of what it should look like and where to look for the needed temperatures.

11. From your graph, find the initial temperature of the water (just before you put the specimen in it; see Figure 1 as an example) and record it.

The initial temperature of the water is _____ °C.

12. From your graph, find the common temperature of the water and specimen after the two have reached the same temperature (where the temperature stops dropping rapidly; see Figure 1 as an example).

The final temperature of the water = final temperature of specimen = _____ °C.

13. Calculate the temperature difference for the water by subtracting the temperature in Procedure 11 from the temperature in Procedure 12.

Specific Heats of Various Materials

The temperature change of the water is _____ C°.

14. Calculate the temperature change of the specimen by subtracting the room temperature in Procedure 1 from the final temperature in Procedure 12.

The temperature change of the specimen is _____ C°.

15. Calculate the heat transferred from the water to the specimen by multiplying the mass (in grams) of the water (Procedure 6), the temperature change of the water in C° (Procedure 13), and the specific heat capacity of water (1 cal/gram-C°).

$Q_{in} = Q_{out}$ = heat transferred from water to specimen = $m_w C_w \Delta T_w$ = _____ cal.

16. Find the specific heat capacity of your specimen by dividing the transferred heat (Procedure 15) by the mass of the specimen (Procedure 4) and the temperature change of the specimen (Procedure 14).

The specific heat capacity = C_{mat} = $\dfrac{Q_{in}}{m_{mat} \Delta T_{mat}}$ = _____ cal/gram-C°.

17. Repeat Procedures 3 through 16 for a second specimen.

SPECIMEN #2 SUMMARY

The mass of the cup is _____ g.

The mass of the specimen is _____ g.

The type of specimen is _____.

The mass of the cup and water is _____ g.

The mass of the water is _____ g.

The initial temperature of the water is _____ °C.

The final temperature of the water and specimen is _____ °C.

The change in the water temperature is _____ C°.

The change in the specimen temperature is _____ C°.

The heat transferred from the water to the specimen is _____ cal.

The specific heat capacity of the specimen is _____ cal/gram-C°.

18. Look up the accepted specific heat capacity values for your two specimens.

The accepted specific heat capacity of Specimen #1 is _____ cal/gram-C°.

The accepted specific heat capacity of Specimen #2 is _____ cal/gram-C°.

19. Find the percentage deviation for each experimental heat capacity by multiplying 100% by the positive difference of the measured value and the accepted value and dividing by the accepted value.

The percentage deviation for Specimen #1 is _____ %.

The percentage deviation for Specimen #2 is _____ %.

➤H. Why is it important that no water splash out of the cup when you add the specimen?

➤I. Why should your results be more accurate when you use Styrofoam cup rather than a regular coffee cup?

➤J.   What are some possible sources of temperature change that are **not** included in the calculations in this exercise? Would these cause your calculated value for the specific heat to be higher or lower than the actual value? Why?

➤K.   Suggest a way to improve the experimental procedure or apparatus to reduce the errors caused by the sources mentioned in Question J.

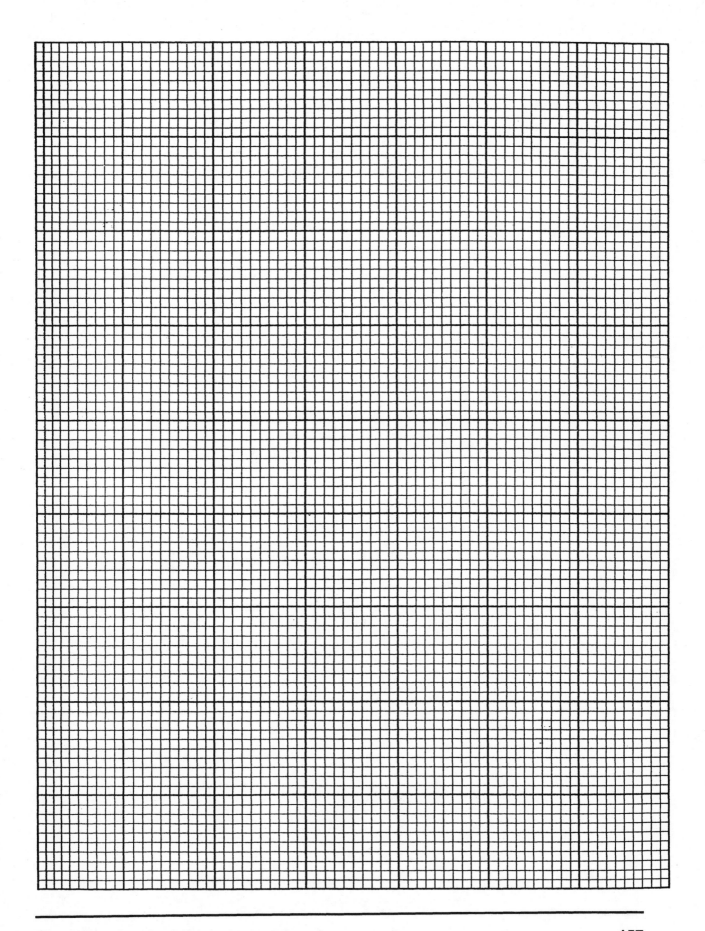

Specific Heats of Various Materials

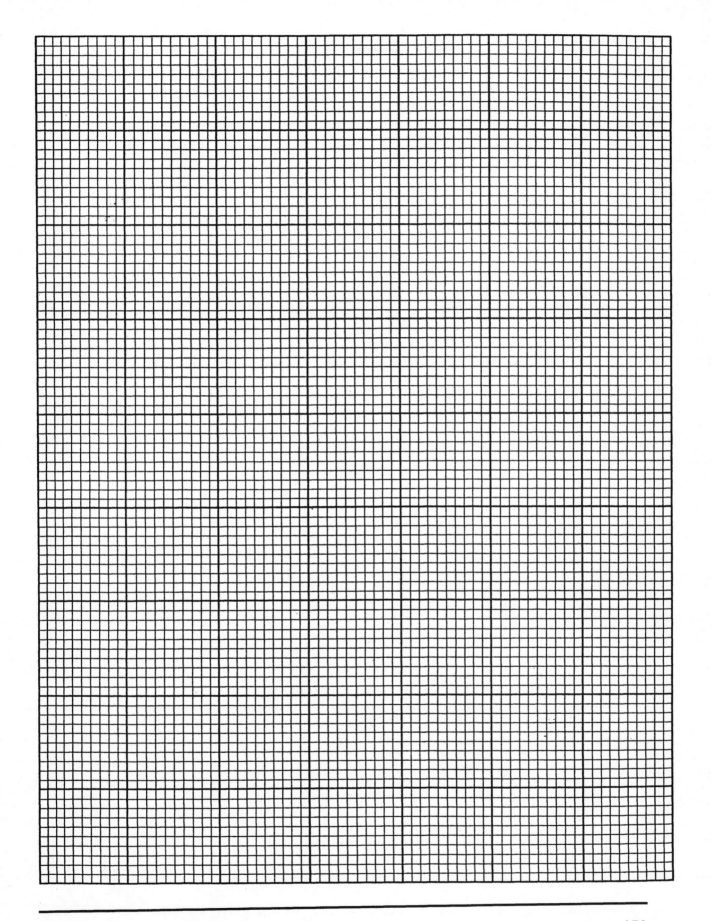

Specific Heats of Various Materials

EXPERIMENT 20

HEAT ENGINES AND REFRIGERATORS

Name_____   Partner(s)_____

Date_____   _____

Section Number_____   _____

BASE CONCEPTS: Temperature, Heat

INTRODUCTION:

Generally speaking, an engine is a device that converts some form of energy into work. In particular, a *heat engine* transforms a flow of thermal energy (heat) into work. In order for a heat engine to operate effectively, it must repeatedly change heat into work (i.e., it must operate in a *cycle*).

Most common engines use thermal energy to change a liquid into a gas. The gas, which takes up a much larger volume than the liquid, pushes on a piston or some other movable object to do work. In order for the cycle to be repeated, either the gas must be exhausted and replaced with new liquid, **or** the gas must be converted back into liquid.

➤A.   If the gas is exhausted from the system, does it take with it any of the energy that was absorbed to make it into a gas? How do you know?

➤B.   If the gas is converted back into a liquid, is any of the energy that was originally absorbed to turn the liquid into a gas not used to do work? How do you know?

In principle, a heat engine can be built with anything that converts heat into work. The general steps are the same. In this lab you will investigate a rubber band heat engine.

SUPPLIES:

Large rubber band, small rubber band, heat gun (or hair dryer), masses.

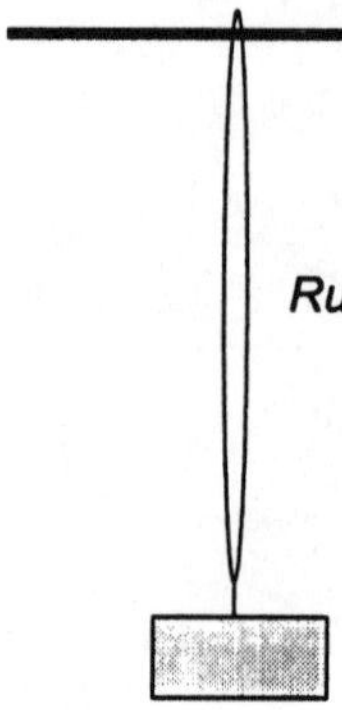

PROCEDURE AND QUESTIONS:

1. Support the large rubber band vertically and hang a mass on the free end of it (see Figure 1). Let the mass hang on the rubber band for a few minutes to make sure the system reaches equilibrium.

Figure 1

2. Heat the rubber band with the heat gun.

Does the mass move? If so, in what direction?

➤C. Did the rubber band absorb heat from the hot air produced by the heat gun? How do you know?

➤D. Was the absorbed heat converted to work done on the masses? How do you know?

Now, suppose you want to make a heat engine with this rubber band. In that case, you will need to leave the mass behind **at the higher level** after it has been lifted by the rubber band. Then, in order to repeat the process, the rubber band will have to return to its initial length.

➤E.   What needs to happen to the temperature of the rubber band in order for the rubber band to return to its original length, that is how must the temperature change if the rubber band is to increase in length?

An engine usually goes through four processes. One process does work, one process absorbs thermal energy, one process expels thermal energy, and a final process is usually needed to get the engine back to its initial state. Consider the rubber band going through such a cycle and think about the order in which the processes occur. The following questions can guide you.

Treat the rubber band as the engine.

➤F.   What do you do to transfer thermal energy to the rubber band?

➤G.   When heat is transferred to the rubber band, how do you know it does work on the mass?

➤H.   What must you do to stop transferring thermal energy to the rubber band?

➤I.   When you stop transferring thermal energy to the rubber band, is the rubber band at a higher or at a lower temperature than the room? How do you know?

➤J.   In order for the rubber band to reach room temperature, must thermal energy be absorbed or removed from the rubber band? How do you know?

➣K. When the rubber band reaches room temperature how should its length compare with its original length?

3. Check with your instructor to make sure your processes are correct before you continue.

➣L. How well would your engine work on a very hot day (one on which the surrounding air has about the same temperature as the heated rubber band)?

Justify your answer.

➣M. Do you expect that the amount of thermal energy transferred into the rubber band will be greater or less than the amount of heat given up by the rubber band when it returns to its original length?

Justify your answer. (*Hint:* Think about conservation of energy.)

➣N. Do you expect that the rubber band can convert **all** the thermal energy it absorbed into work? (If this could happen, from conservation of energy, no thermal energy would be transferred from the rubber band after the mass was lifted and removed, which is to say the temperature of the rubber band after lifting the mass would be room temperature.)

Justify your answer.

THE RUBBER BAND REFRIGERATOR

Think about using the rubber band as a refrigerator. First answer the following questions, and discuss the answers with your group and instructor.

➤L.   Does the temperature of a rubber band go up or down when it is stretched rapidly? To test this, use your upper lip as a thermometer, by resting the rubber band there until it feels the same temperature as your lip. Then stretch it rapidly and see if it feels warmer or cooler right after stretching.

In this process, what did you do to cause the temperature to change?

➤M.   What happens to the temperature of a stretched rubber band when it is allowed to rapidly contract? You can find out by taking an already stretched rubber band and resting it on your lower lip. When it feels the same temperature as your lip, let it contract rapidly and see if it feels cooler or warmer to your lip.

In this process, what did you do to cause the temperature to change?

Suppose you have access to two rooms that are both thermally insulated from the outside world and from each other. Let Room C be the room you want to cool and Room H be the other room. Also assume that both rooms start at the same temperature.

➤N.   Will the rubber band absorb heat from its surroundings when it is at a lower or at a higher temperature than the surroundings?

➤O.   In which room do you want heat absorbed?

Why?

➤P.   When you stretch a rubber band in a room (suppose they both start at the same temperature) will the rubber band absorb or give off heat after it is stretched?

How do you know?

➤Q.   When you take a stretched rubber band that is in a room at the same temperature and let it go back to its original length, will the rubber band absorb or give off heat after it has returned to its original length?

How do you know?

Using your answers to N–Q as justifications, answer the following questions.

➤R.   In what room would you stretch the rubber band?

Why?

➤S. After stretching the rubber band, would you stay in the room in which you stretched it for a while, or would you leave immediately? Justify your answer

➤T. After leaving the room and moving to the other room, what would you do to the rubber band?

Why?

➤U. After performing the action of Question T, would you leave the room after a while or immediately? Justify your answer.

➤V. If you wanted this to be a cycle, what would you do next?

➤W.   Write your cycle down with the following steps justified (how and in what room):

1) Heat absorbed

2) Heat emitted

3) Work done **on** refrigerator

This information can be used to form an analogy with your home refrigerator. The inside of the refrigerator is the "cooled room"; the kitchen is the thermally isolated "room next door"; and the electricity is your "stretching" the rubber band.

EXPERIMENT 21   ENERGY AWARENESS

Name_____   Partner(s)_____

Date_____   _____

Section Number_____   _____

BASE CONCEPTS: Forms of Energy, Energy Resources

INTRODUCTION:

Energy resources and their utilization are major subjects of concern and debate in the world today. Something that can be viewed as an abstract scientific concept drives much of our social and economic life. This lab will help you recognize more clearly the central role energy and energy resources play in your daily life.

PROCEDURE:

To increase your awareness of the importance of energy in our society, you will keep track of situations in which you use some form of energy (other than your own) and situations in which you choose **not** to use some form of energy (other than your own) or to use it more efficiently.

Your instructor will tell you what days should be covered by this energy audit. You should use the following three categories to organize the information you record.

1.   *Direct* energy uses: These are uses for which the energy comes directly to you from the energy sources, such as lighting, heat, and motor-driven devices.

2.   *Indirect* uses: These are uses for which the energy comes indirectly from the energy sources. An example is reading a newspaper. Although *you* may not explicitly use energy to obtain or read the newspaper, energy was used in cutting down trees, making paper from the trees, transporting the paper, printing the newspaper, and delivering it.

3.   *Conservation*: Here you record situations in which you chose **not** to use an energy source other than yourself or in which you arranged your activities to make more efficient use of an energy source.

For each item in these three categories, you are to clearly state what you did to use or not to use energy and *why* this activity used or saved energy, indicating the type of energy resource involved. A *partial* sample of an audit for one day is given on p. 170.

Friday's Audit

Direct Uses

Lights on in living area — electricity used to make light.
Drove car to class — used gasoline.
Heated room — used natural gas to provide heat to room.
Washed clothes in washing machine — used electricity and water.
Rode in car to movie — used gasoline.
Drove car to restaurant — used gasoline.

Indirect Uses

Ate bacon for breakfast — energy for pig's growth/cooking/transportation.
Ate at restaurant — energy for cooking, washing dishes, food transportation.
Read and threw away newspaper — energy to make and deliver newspaper.
Threw out a burned-out light bulb — energy to make and deliver bulb.

Conservation

Turned down heat, wore a sweater — saved gas/electricity for heat.
Walked to school instead of driving — saved gasoline for car.
Used *my* energy to recycle newspaper — saved part of a tree.

Note that the first two categories are not necessarily distinct, but you should record a particular use in **only one** of the two groups. For example, if you cooked eggs for breakfast, you could enter either "used stove to cook eggs" as a direct use **or** "ate cooked eggs" as an indirect source.

In addition to your energy audit, you are to write a paper (about 2 pages in length) discussing ways you could have reduced the energy you used during the audit period and the impact they might have on your life. The essay should also include ways that society could help in reducing energy use (e.g., providing a good bus system in your community).

EXPERIMENT 22

HOW MUCH ENERGY IS SAVED BY TURNING DOWN THE THERMOSTAT?

Name_____   Partner(s)_____

Date_____   _____

Section Number_____   _____

BASE CONCEPTS: Energy, Temperature, Heat

INTRODUCTION:

The claim is frequently made that a substantial amount of energy could be saved if people turned their home thermostats down from 75°F to 65°F across the country. In this laboratory you will investigate the percentage of energy that could be saved as a result of such a change.

The amount of energy lost in a given time by a system (e.g., a house) to its surroundings (e.g., the outdoors) is approximately proportional to the *temperature difference* between the system and its surroundings. Since it will be easiest for you to perform the experiment at room temperature, some numbers need to be calculated.

Let's assume that a typical winter outside temperature is 30°F = – 1.1°C. A warm house at 75°F = 23.9°C needs to have a certain amount of heat added per unit time (e.g., every hour) to maintain the temperature difference between it and the outside. If, instead, the thermostat is set at 65°F = 18.3°C, less heat is required. For these values, the temperature difference between the warm house and the outside is 23.9°C –(– 1.1°C) = 25.0°C. The corresponding difference for the cooler house is 18.3°C –(– 1.1°C) = 19.4°C.

SUPPLIES:

Lightbulb in socket, thermometer, timer, clock, large container.

PROCEDURE:

1. Obtain a lightbulb (about 50 W) with a known power output, a large wide-mouth container that the lightbulb will fit into, some insulating material, a computerized temperature probe, and a timer that can be stopped and restarted without rezeroing.

---

2. Using your thermometer, determine the temperature of the room in which you are working.

   The room temperature is _____ °C.

3. Place the lightbulb and the thermometer inside the container, now known as your "house." Position the thermometer so that it reads the air temperature in the house.

4. Compute the temperature that is the same number of Celsius degrees above your room temperature (found in Procedure 2) as the *warm house* temperature is above a typical outside temperature. Do this by adding 25.0°C to your measured room temperature.

   The target *warm* temperature = room temperature + 25.0°C is _____ °C.

5. Repeat Procedure 4 for the *cooler house* by adding 19.4°C to your room temperature.

   The target *cooler* temperature = room temperature + 19.4°C = _____ °C.

6. Designate three members of your group to be (1) the thermostat, (2) the timekeeper, and (3) the thermometer reader. The thermostat is responsible for turning the light on and off. The timekeeper keeps a running total of the time that the light is on. The thermometer reader monitors the temperature and lets the thermostat know when to turn the light on or off.

7. First, you need to get the house up to the desired temperature. The timekeeper should make sure that the timer is zeroed before you start. The thermostat turns on the light until the thermometer reads 1°C above the target temperature for the **warm** house. (Don't start the timer yet!)

8. From a clock in the room, note the time at which the house reaches this temperature.

   The starting time is _____ hour _____ minutes _____ seconds.

9. Turn off the light in the house.

10. When the house temperature reaches 1°C *below* the **warm** target temperature, the thermometer reader tells the group. The thermostat turns *on* the light. At the same time, the timekeeper pushes the *start* button on the timer to keep track of the time elapsed when the light is on.

11. When the house temperature reaches 1°C *above* the **warm** target temperature, the thermometer reader tells the group. The thermostat turns *off* the light. At the same time, the timekeeper pushes the *stop* button on the timer to halt recording of the time elapsed when the light is on. Since the timekeeper should be keeping a running total of the time the light is on, be sure **not** to rezero the timer.

12. Repeat Procedures 10 and 11 for exactly 15 minutes after the recorded starting time. After you have stopped, record the total amount of time that the light was on, as indicated by the timer.

    The total time the light was on is _____ seconds.

13. In order to do the second part of the experiment, you must make sure that your system starts at the **cooler** target temperature. While you let things cool down, one person should carefully monitor the thermometer while the others do the following calculation.

    Calculate the energy used in kilowatt hours (kWh).

    1 kW (kilowatt) = 1000 W (watt).
    1 hour = 3600 seconds.

    The time *in hours* the light was on = time in seconds × (1 hour/3600 seconds) = _____ hours.

    The lightbulb power *in kilowatts* = power in watts × (1 kW/1000 W) = _____ kW.

    The energy *in kWh* used = (time in hours) × (power in kilowatts) = _____ kWh.

14. Once the system is cooled to 1°C above the target temperature for the **cooler** house, write down your starting time and repeat Procedures 10 through 12 using the **cooler** target temperature found in Procedure 5. Make sure that the elapsed time between first reaching this temperature and stopping the experiment is 15 minutes, the same amount of time as the first experiment.

    The starting time for the house at the cooler target temperature is

    _____ hour _____ minutes _____ seconds.

    The total time the light was on is _____ seconds = _____ hours.

    The energy used is _____ kWh.

15.   To compute the percentage of energy saved, first find the difference in the two energies, then divide by the energy used for the warm house, and multiply by 100%:

$$\% \text{ ENERGY SAVED} = \frac{\text{ENERGY DIFFERENCE}}{\text{WARM HOUSE ENERGY}} \times 100\%.$$

The percentage of energy saved is _____ %.

GLOBAL QUESTIONS:

A.   What else can you think of that could reduce the amount of heat used in winter?

B.   Gas energy costs less than electrical energy. Why might this be true?

C.   Were you surprised by the percentage of energy saved?   Why or why not?

D.   How might our original assumption of 30°F = −1.1°C for a typical winter temperature influence the percentage of energy saved?

EXPERIMENT 23

THERMAL INSULATORS

Name_____   Partner(s)_____

Date_____   _____

Section Number_____   _____

BASE CONCEPTS: Temperature, Heat, Graphing

INTRODUCTION:

Different materials have varied abilities to support a temperature difference between two regions. The longer it takes for the two regions to reach the same temperature, the better the insulation. The time required for the temperatures to become equal depends on many factors — the initial temperature difference, the distance between the two regions, the cross sectional area of the regions, and the type of material between the regions.

In this lab, the area, thickness, and initial temperature difference will all be held constant. Only the type of material between the regions will be changed.

SUPPLIES:

Soda cans with tops cut off, 6 oz juice cans with lids removed, computerized temperature probes, hot water reservoir, various materials (e.g., housing insulation, gravel, air, rubber bands, crushed paper, cotton), hair dryer or heat gun.

PROCEDURE:

1.   Choose three different insulators with which you would like to work and enter their names as column headings in Table 1 on p. 178.

2.   Make a mark on the side of the juice can about one-fourth of the way down from the top. Each time you fill the can, make sure you fill it to this specific height.

3.   Fill the juice can to the mark with hot water. Measure and record its temperature. (In Procedure 6 you will want to make sure your heated insulation is within 1° of this temperature.) Place the juice can inside the soda can without any added insulation. Place your thermometer in the water. Begin to record the temperature of the water.

The measured hot water temperatures are _____ °C and _____ °C.

Thermal Insulators   **177**

Table 1
Insulation Data

INSULATION MATERIAL→	NO ADDED INSULATION			
TIME (min)	TEMPERATURE (°C)	TEMPERATURE (°C)	TEMPERATURE (°C)	TEMPERATURE (°C)
0				
4				
8				
12				
16				
20				

4.   Wait until the thermometer reads a few degrees less than the initial temperature of the water. This temperature will be used as the "starting temperature" for all the trials.

5.   Restart the readings of the temperature probe. The "time = 0 minutes" entry should be the starting point. Record data for 20 minutes and then read the values off the computer graph or data table for the temperature readings every 4 minutes.

6.   Remove the juice can. Dry the inside of the soda can if necessary. Place about ½ in. of one of your chosen insulating materials in the bottom of the soda can. Heat up the insulating material with the heat gun or hair dryer until it is the desired temperature (within 1° of the hot water temperature determined in Procedure 3). Replace the water in the juice can with hot water, again up to the mark. Place the juice can inside the soda can and pack some insulating material in the space between the two cans. When the temperature of the water reaches the same "starting temperature" as in Procedure 4, begin to take data and fill in the appropriate column in Table 1.

7.   Repeat Procedure 6 for the other two insulators.

8.   Plot your data on a graph of temperature (°C) versus time (minutes). Use different symbols for the different insulating materials (different data columns). When appropriate, use this graph to help you answer the questions on p. 179.

           Thermal Insulators

GLOBAL QUESTIONS:

A.   What does it mean to be a good insulator?

B.   Which of your three materials was the best insulator?

How do you know?

C.   Which of your three materials was the worst insulator?

How do you know?

D.   Why did you have to start all the data collection trials at the same starting temperature?

E.   Why did you collect data with no added insulator?

F.   Why did you need to heat the insulation material so that its temperature was near that of the
      hot water?

G.   How could this lab be improved?

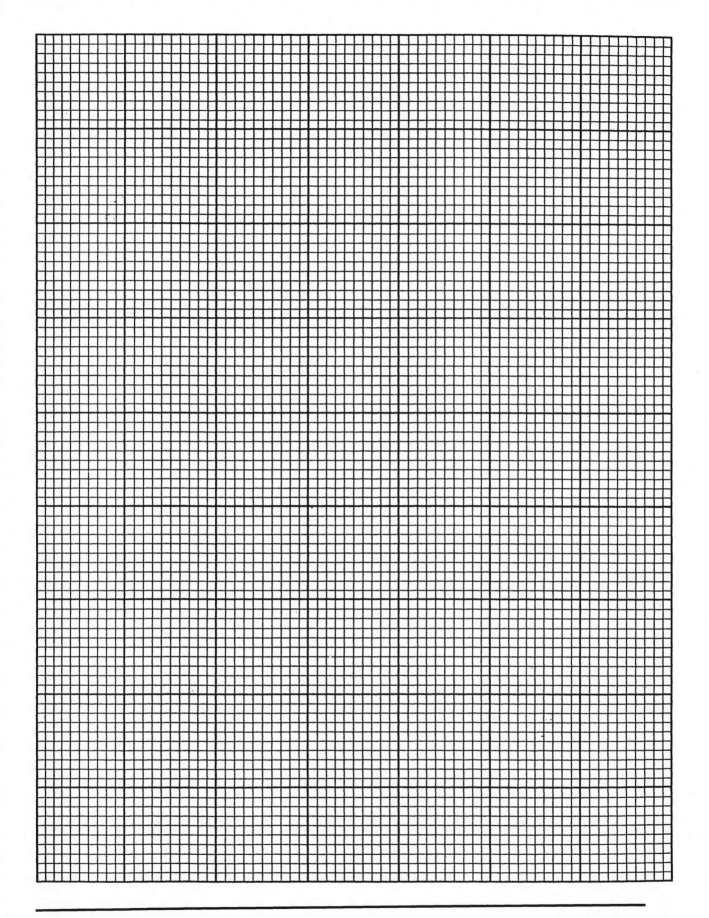

EXPERIMENT
24

WAVES AND
WAVE CHARACTERISTICS

Name_____   Partner(s)_____

Date_____   _____

Section Number_____   _____

BASE CONCEPTS: Work, Energy, Waves

INTRODUCTION:

A *wave* is a traveling disturbance that transfers energy from one place to another without transferring matter. When the disturbance is rhythmic (or repetitive), it is called a *periodic wave*. For a *longitudinal* wave (e.g., sound) the disturbance is in the same direction as the energy transfer. For a *transverse* wave (e.g., light) the disturbance is perpendicular to the energy transfer direction.

In this laboratory you will investigate four properties of a transverse periodic wave on a rope. Before starting, it is necessary to clarify some terms that are used in discussing periodic waves. This will be done in the context of a wave traveling along a rope. Figures 1 and 2 will help you to understand these terms.

A snapshot of a wave on a rope
Time is a set value

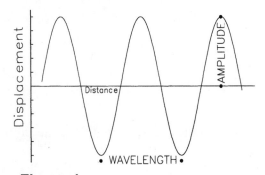

Figure 1

Displacement of a string at a point on the
string as a function of time

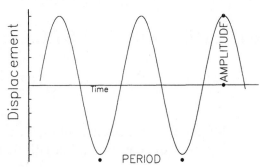

Figure 2

Characteristics of a Wave   **185**

DISTURBANCE: The *disturbance* is the occurrence that is repeated. If the rope is disturbed by moving it repeatedly up and down, the disturbance will consist of the motion upward from the initial (equilibrium) position, the motion downward past the initial position, and then the motion upward from the lowest position to the initial position again. At this point, the disturbance will be repeated.

AMPLITUDE: The *amplitude* is the maximum distance the rope moves from its initial position.

WAVELENGTH: The *wavelength* is the distance along the rope between the beginning of one disturbance and the adjacent repetition at some point in time. It is also the distance along the rope between the maximum displacement in one direction and the adjacent maximum displacement in the same direction (or between **any** two successive positions that have identical displacements and slopes).

PERIOD: The *period* is the time required for a single complete disturbance pattern to pass a given point. Equivalently, it is the shortest time in which the motion of one point on the rope will repeat itself.

FREQUENCY: The *frequency* is the number of disturbances that pass a point along the rope per time interval. For example, if you move the rope up and down to produce 5 repetitions of the disturbance in 2 seconds, the frequency will be 5 repetitions/2 seconds = 2.5 repetitions/seconds (frequently called cycles/second — a pun!). Rather than writing the units as repetitions/second, a standard name is defined: 1 repetition/second = 1 hertz (Hz). (Note that frequency is the reciprocal of period, which is the time/repetition).

SPEED: The *speed* of the wave is the distance some feature of the disturbance (e.g., a peak or a valley) moves along the rope per time interval. Note you can calculate the speed if you know the wavelength and the period. Since the wave travels one wavelength in one period, the speed is just wavelength/period. This can also be written as wavelength x frequency.

SUPPLIES: Meterstick, tape, timer, 6-m-long rope.

PROCEDURE AND QUESTIONS:

1. Tie one end of the rope to some stationary object near the floor.

2. Stretch the rope out in a straight line and mark this line on the floor. This line is the initial position (or equilibrium line) of the rope.

3. Produce a wave by moving the free end of the rope rhythmically back and forth horizontally across the floor.

➢A.   Are you producing a longitudinal or transverse wave? Justify your answer from the definitions given at the beginning of this lab.

4.   Try to increase the amplitude.

➢B.   Does it feel like you are doing more or less work?

What are you doing differently to produce the increased amplitude?

➢C.   You are doing work on the rope because you are exerting a force in the direction of the rope's motion. When you do work on the rope, you transfer energy to it. Where does this energy go?

➢D.   When you move your hand farther back and forth on the floor (i.e., increase the amplitude of the wave), are you doing more or less work on the rope?

How do you know?

Characteristics of a Wave   **187**

5. While one lab partner keeps time, count the number of repetitions you produce in 15 seconds. When the 15 seconds is up, drop the free end of the rope and let the rope's wave formation remain on the floor.

The number of repetitions in 15 seconds is _____.

The frequency = number of repetitions/15 seconds = _____ Hz.

6. You can find the wavelength by measuring the distance between the same points on two adjacent repetitions on the floor. When you let the free end go, the disturbances should still be exhibited on the rope.

The wavelength is _____ cm.

7. The speed can be calculated by multiplying the wavelength and frequency together.

The speed of the wave is _____ cm/s.

GLOBAL QUESTIONS:

E. Would the speed of the wave increase, decrease, or remain the same if you increased the repetitions/time at your end of the rope? Why?

F.   Would the speed increase, decrease, or remain the same if, leaving the repetitions/time the same, you increased the amplitude? Why?

G.   Write down three types of waves that you know about. For each, describe how you could measure one of the characteristics defined at the beginning of this lab.

EXPERIMENT
25   REFLECTION OF LIGHT

Name_____   Partner(s)_____

Date_____   _____

Section Number_____   _____

BASE CONCEPTS: Light Ray, Law of Reflection

INITIAL QUESTION:

A.   You walk into a room that has a mirror covering the wall in front of you. You are facing the mirror and are unaware that the mirror is there. A stranger walks into the room behind you and to your right. Where do you see the stranger? (You may indicate where you see the stranger in Figure 1.) Where do you see yourself? (You may indicate where you see yourself in Figure 1.) Why did you choose these locations?

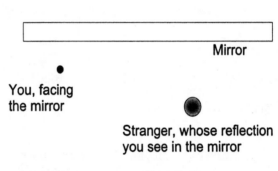

Mirror

You, facing
the mirror

Stranger, whose reflection
you see in the mirror

Figure 1

INTRODUCTION:

When investigating reflection from a smooth surface, it is useful to think of light as traveling in a straight line unless it encounters an object in its path. In physics these straight lines are called *light rays*. Your brain assumes that light rays that reach your eyes have traveled in a straight line. You therefore "see" an object at the point from which these light rays *seem to diverge*. If there have been no reflections (or refractions), then your brain "sees" the object where it actually exists. If, however, the light from the object has been reflected from a smooth surface, your brain still processes the light in the same manner. You "see" the object at the point from which the light appears to have diverged.

In this lab, you will investigate reflections from plane mirrors. In order to understand your results, you need to recall the *law of reflection*. A light ray traveling from its source toward the mirror is called an <u>incident ray</u>. A ray that has undergone reflection is called a <u>reflected ray</u>. Assume that the ray has been reflected at some point A on the mirror. The law of reflection states that the angle the incident ray makes with a line that intersects the mirror perpendicularly at point A (the normal) is equal to the angle the reflected ray makes with the same normal (see Figure 2 below). This law can be cumbersome to use when the mirror is curved. For a plane mirror though, since the normals are parallel for all points on the mirror surface, it is much easier to use.

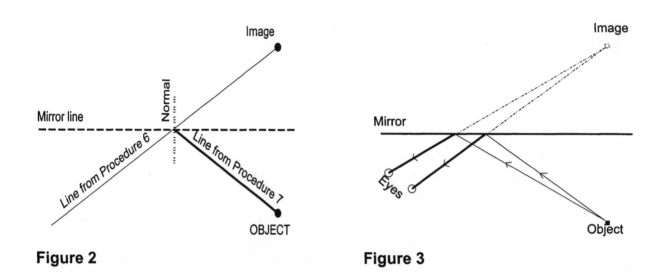

Figure 2   **Figure 3**

As an example, look at Figure 3 showing reflected rays from the person who walked in behind you in the mirrored room. Only two rays that emanate from the stranger and enter your eyes are drawn. The thin solid lines are the incident rays, the thick solid lines are the reflected rays, and the dashed lines are extensions of the reflected rays to show the point from which your eyes "see" the rays as diverging.

SUPPLIES:

Plane mirror (about ½ in. high with no border), pins, ruler, paper, protractor, block of wood.

PROCEDURE:

1.   On a white piece of paper taped to cardboard, draw a straight line slightly longer than your mirror.

2.   With the mirror mounted on the front of a block of wood, making sure the reflective part is along one edge of the block, place the mirror along the line drawn in Procedure 1.

3.   Stick a tall straight pin into the cardboard about 10–12 cm in front of the mirror and toward the right side (see Figure 4).

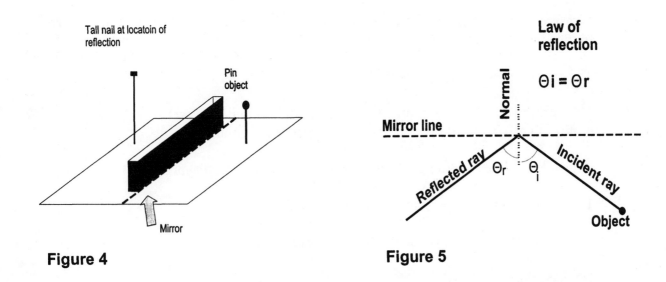

Figure 4

Figure 5

4. Place a large nail, taller than the mirror, behind the mirror where the image of the pin **appears** to be. To check the location, move your head around. If you have placed the nail in the correct position, then when viewed from any angle the nail and the reflection should appear to be at the same location.

5. After securing the nail in the location you found for the image in Procedure 4, remove the mirror and piece of wood, leaving the straight pin, the large nail, and the line where the mirror was located.

6. Draw a straight line on the paper that starts at the point where the nail is located and ends some place on the side of the mirror line where the pin (object) is located (see Figure 5). The portion of this line on the object side of the mirror line represents the reflected ray.

7. In a different color, draw a line that starts at the point on the paper where the pin (object) is located and ends at the point where the line drawn in Procedure 6 intersects the mirror line. This line represents the incident ray.

8. Draw a line (the *normal*) perpendicular to the mirror line and located at the point where the mirror line intersects the lines drawn in Procedures 6 and 7 (see Figure 5).

9. Measure and record the angle between the incident ray (drawn in Procedure 7) and the normal (drawn in Procedure 8). This is the incident angle (Θ_i).

The angle the incident ray makes with the normal to the mirror is _____°.

10. Measure and record the angle between the reflected ray (drawn in Procedure 6) and the normal (drawn in Procedure 8). This is the reflected angle (Θ_r).

The angle the reflected ray makes with the normal to the mirror is _____ °.

11. Compare the values in Procedures 9 and 10. Are they within 2° of each other? _____ (yes or no) If not, check with your lab instructor before continuing.

12. Repeat Procedures 6 through 11, drawing a different line starting at the nail.

The angle the incident ray makes with the normal is _____ °.

The angle the reflected ray makes with the normal is _____ °.

Do these differ by less than 2°? _____ (yes or no)

GLOBAL QUESTIONS:

B. If you can see the reflection of the eyes of a person in a mirror, can that person see your eyes? How do you know?

C. Why were you not asked to find a percent difference in this lab?

D. Draw a line that is perpendicular to the mirror line and goes through the position of the object. This is the object line. Draw a line perpendicular to the mirror line that goes through the apparent position of the image. This is the image line. How long is the object line as measured from the object to the intersection with the mirror line?

E. How long is the image line as measured from the image to the intersection with the mirror line?

F. How far to the right of the left edge of the mirror did you place the object (pin)?

G. How far to the right of the left edge of the mirror does the object line intersect the mirror line?

H.   From this information (D–G) how could you generally figure out where you will see an object reflected by a plane mirror?

Bumper sticker on the back of a large tanker truck: *"If you can't see my mirrors, I can't see you."*

EXPERIMENT 26

SERIES ELECTRICAL CIRCUITS

Name_____   Partner(s)_____

Date_____   _____

Section Number_____   _____

BASE CONCEPTS: Current, Potential Difference, Resistance, Ohm's Law

INTRODUCTION:

The potential drop across any Ohmic resistor is equal to its electrical resistance times the current passing through the resistor; i.e., $V = IR$, where V is the potential drop in volts (V), I is the current in amps (A), and R is a constant resistance in ohms (Ω). This relation is commonly known as <u>Ohm's law</u>. In any circuit where there is a single power supply, it is convenient to define an *equivalent resistance* for the circuit. This is the value of the resistor that could replace the entire circuit and leave the operating parameters of the power supply (potential difference across it and current flowing from the power supply) unchanged. Experimentally, then, the equivalent resistance is just the potential difference across the power supply divided by the current flowing through it.

Two resistors can be connected in a circuit in just two ways (see Figure 1). (Of course, circuits can become much more complicated when there are more than two elements.) In a parallel circuit, there is more than one path through which the current can flow. In a *series circuit*, there is only one path through which the current can follow. In this case, the current passing through each of the elements in the circuit is the same. In this lab you will study a two-resistor series circuit.

SUPPLIES:

Multimeter, 100–200 Ω resistor, 500--1000 Ω resistor, 12–20 V direct current power supply, connecting wires with alligator clips on each end.

PROCEDURE:

1.   Set the multimeter to function as an ohmmeter by selecting "Ohms" on the selector dial.

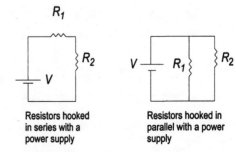

Resistors hooked in series with a power supply

Resistors hooked in parallel with a power supply

Figure 1

2.   Measure and record the values of both of your resistors. Let R_1 be the smaller of the two

---

resistance values. Note that the multimeter measures either in Ω or kΩ (1 kΩ = 1000 Ω). If the measurement is in kΩ, you need to move the decimal point to the right three places (i.e., multiply the reading by 1000).

$R_1 =$ _____ Ω.

$R_2 =$ _____ Ω.

3. Using three wires with alligator clips on each end, complete the series circuit by connecting the following (see Figure 2):

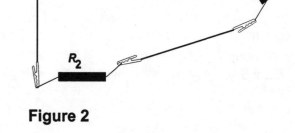
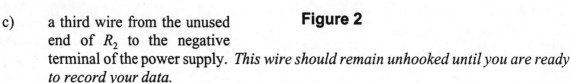

a) one wire from the power supply positive terminal to one end of R_1,

b) a second wire from the unused end of R_1 to one end of R_2, and

c) a third wire from the unused end of R_2 to the negative terminal of the power supply. *This wire should remain unhooked until you are ready to record your data.*

Figure 2

4. Now, set your multimeter to function as a voltmeter by turning the selection dial to "DC Volts." When measuring voltages, you must always connect the voltmeter **across** (in parallel with) the component of interest (see Figure 3).

5. After checking with your instructor, connect the wire to the negative terminal of the power supply. Set the voltage on the power supply to 10 V.

6. Measure the voltage drops (i.e., potential difference) across each of the two resistors and across the power supply.

The voltage drop across R_1 (put multimeter probes on each side of R_1) is $V_1 =$ _____ V.

The voltage drop across R_2 (put multimter probes on each side of R_2) is $V_2 =$ _____ V.

The voltage drop across the power supply (put multimter probes on the two power supply outputs) is _____ V.

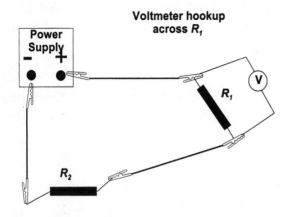

Figure 3

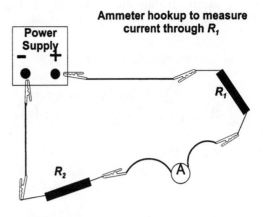

Figure 4

7. Now, set your multimeter to function as an ammeter by choosing "DC Current" on the selection dial. When measuring current, you must hook the ammeter in series with the appropriate component (see Figure 4). Also, it is easy to blow a fuse inside the multimeter when you are using it as an ammeter. To avoid doing this, always **turn the measuring scale setting to the maximum value before connecting the ammeter**. When taking your readings, carefully turn the scale down until you have the most accurate reading possible.

8. Connect the ammeter between the positive terminal and R_1 by

    a) disconnecting the alligator clip from the positive terminal of the power supply and reconnecting it to the negative side of the ammeter, and

    b) connecting a wire between the positive side of the ammeter and the positive terminal of the power supply.

    The reading from the ammeter connected in this way gives you the current flowing through the power supply. If the current is in mA (1 mA = 0.001 A), you must move the decimal point three places to the left (i.e., multiply by 0.001) to obtain your result in amps.

    The current through the power supply is _____ A.

9. First, reconnect the circuit as it was before starting Procedure 7. Then, connect the ammeter between R_1 and R_2 in the same manner as in Procedure 7 (see Figure 4). Read and record the current through R_1.

    The current through R_1 is $I_1 =$ _____ A.

10. Again return the circuit to its original form, then place the ammeter between R_2 and the negative side of the power supply. Read and record the current through R_2.

The current through R_2 is I_2 = _____ A.

11. To verify Ohm's law, calculate the voltage drop across each resistor divided by the current through it. The unit for V/A is Ω (i.e., 1 V/A = 1 Ω).

V_1/I_1 = _____ Ω.

V_2/I_2 = _____ Ω.

12. Compute the percent deviation between the values calculated in Procedure 11 and the resistances R_1 and R_2 measured in Procedure 2. Treat the values from Procedure 2 as the "accepted values." (Recall that the percent deviation = the positive difference between the experimental and accepted values divided by the accepted value times 100%).

The percent deviation between R_1 and V_1/I_1 is _____ %.

The percent deviation between R_2 and V_2/I_2 is _____ %.

13. Calculate the equivalent resistance of the circuit by dividing the potential difference across the power supply (from Procedure 6) by the current through the power supply (from Procedure 8).

The equivalent resistance is R_{eq} = _____ Ω.

14. Sum R_1 and R_2 (from Procedure 2) and compute the percent deviation between this sum and the equivalent resistance. Treat the sum as the "accepted value."

The percent deviation between $R_1 + R_2$ and R_{eq} is _____ %.

15. Sum the voltage drops across the two resistors (from Procedure 6) and compute the percent deviation between this sum and the potential drop across the power supply (also from Procedure 6). Treat the power supply potential drop as the "accepted value."

The percent deviation between $V_1 + V_2$ and the power supply drop is _____ %.

GLOBAL QUESTIONS:

A.   Was the potential drop larger across the smaller or across the larger resistor?

Explain why this makes sense in terms of Ohm's law if the current through each component is the same.

B.   Lightbulbs are made of fine wires inside an evacuated glass bulb. When a lightbulb burns out, the wire breaks, leaving no path for the current. If you replaced the two resistors you used with lightbulbs in this series circuit, would the good lightbulb remain lit if the other lightbulb burned out?

Why or why not?

C.   Suppose you hooked a third resistor (R_3) in series with the two resistors you used in this experiment.

In terms of R_1, R_2, and R_3 what is the equivalent resistance of the circuit? (*Hint:* Look at Procedure 14.)

D.   Would the current from the power supply be higher or lower with the three resistors than it was with the two?

Why? *(Hint:* Use your answer from Question C and the definition of equivalent resistance to justify your answer.)

E.   In size, how does R_{eq} compare with R_1 and R_2? Is it larger than either of them, smaller than either of them, or somewhere between the two values?

EXPERIMENT
27

PARALLEL ELECTRICAL CIRCUITS

Name_____   Partner(s)_____

Date_____   _____

Section Number_____   _____

BASE CONCEPTS: Current, Potential Difference, Resistance, Ohm's Law

INTRODUCTION:

The potential drop across any Ohmic resistor is equal to its electrical resistance times the current passing through the resistor; i.e., $V = IR$, where V is the potential drop in volts (V), I is the current in amps (A), and R is a constant resistance in ohms (Ω). This relation is commonly known as *Ohm's law*. In any circuit where there is a single power supply, it is convenient to define an *equivalent resistance* for the circuit. This is the value of the resistor that could replace the entire circuit and leave the operating parameters of the power supply (potential difference across it and current flowing through it) unchanged. Experimentally, then, the equivalent resistance is just the potential difference across the power supply divided by the current flowing through it.

Two resistors can be connected in a circuit in just two ways (see Figure 1). (Of course, circuits can become much more complicated when there are more than two elements.) In a series circuit, there is only one path through which the current can follow. In this case, the current passing through each of the elements in the circuit is the same. In a *parallel circuit*, there is more than one path through which the current can flow. In this lab you will study a two-resistor parallel circuit.

SUPPLIES:

Multimeter, 100–200 Ω resistor, 500–1000 Ω resistor, 12–20 V power supply, connecting wires with alligator clips on each end.

PROCEDURE:

1.   Set the multimeter to function as an ohmmeter by selecting "Ohms" on the selector dial.

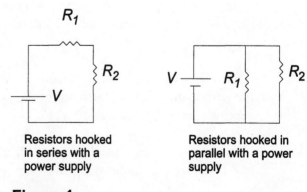

Resistors hooked in series with a power supply

Resistors hooked in parallel with a power supply

Figure 1

---

2. Measure and record the values of both of your resistors. Let R_1 be the smaller of the two resistance values. Note that the multimeter measures either in Ω or $k\Omega$ ($1\ k\Omega = 1000\ \Omega$). If the measurement is in $k\Omega$, you need to move the decimal point to the right three places (i.e., multiply the reading by 1000).

$R_1 = $ _____ Ω.

$R_2 = $ _____ Ω.

3. Using four wires with alligator clips at each end, complete the parallel circuit by connecting the following (see Figure 2):

    a)   one wire from the positive power supply terminal to one end of R_1,

    b)   a second wire from the same end of R_1 to one end of R_2,

    c)   a third wire from the unused end of R_2 to the unused end of R_1, and

    d)   a fourth wire between the end of R_1 to which the third wire was just connected and the negative terminal of the power supply. *This wire should remain unhooked until you are ready to record your data.*

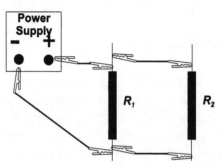

Two Resistors Hooked in Parallel

Figure 2

4. Now, set your multimeter to function as a voltmeter by turning the selection dial to "DC Volts." When measuring voltages, you must always connect the voltmeter **across** (in parallel with) the component of interest (see Figure 3).

5. After checking with your instructor, connect the wire to the negative terminal of the power supply. Set the voltage to 10 V.

6. Measure the voltage drop across each of the two resistors and across the power supply.

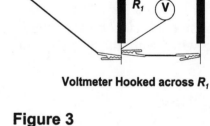

Voltmeter Hooked across R_1

Figure 3

The voltage drop across R_1 is _____ V.

The voltage drop across R_2 is _____ V.

The voltage drop across the power supply is _____ V.

7. Now, set your multimeter to function as an ammeter by choosing "DC Current" on the selection dial. When measuring current, you must hook the ammeter in series with the appropriate component (see Figure 4). Also, it is easy to blow a fuse inside the multimeter when you are using it as an ammeter. To avoid doing this, always **turn the measuring scale setting to the maximum value before connecting the ammeter**. When taking your readings, carefully turn the scale down until you have the most accurate reading possible.

8. Connect the ammeter between the power supply positive terminal and R_1 by

   a) disconnecting the alligator clip from the positive terminal of the power supply and reconnecting it to the negative side of the ammeter, and

   b) connecting a wire between the positive side of the ammeter and the positive terminal of the power supply.

   The reading from the ammeter connected in this way gives you the value of the current flowing through the power supply. If the current is in mA (1 mA = 0.001 A), you must move the decimal point three places to the left (i.e., multiply by 0.001) to obtain your result in amps.

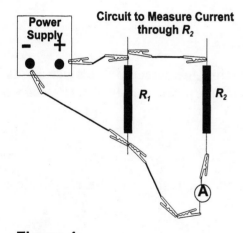

Figure 4

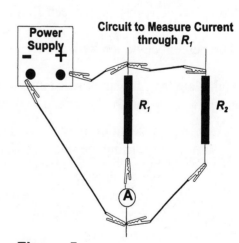

Figure 5

The current through the power supply is _____ A.

9. First, reconnect the circuit as it was before starting Procedure 7. Then connect the ammeter in series with R_1 to measure the current flowing through it (see Figure 4). Read and record the current through R_1.

The current flowing through R_1 is $I_1 =$ _____ A.

10. Again return the circuit to original form, then connect the ammeter to measure the current through R_2 as shown in Figure 5. Read and record the current through R_2.

The current flowing through R_2 is $I_2 =$ _____ A.

11. To verify Ohm's law, calculate the voltage drop across each resistor divided by the current through it. The unit for V/A is Ω (i.e., 1 V/A = 1 Ω).

$V_1/I_1 =$ _____ Ω.

$V_2/I_2 =$ _____ Ω.

12. Compute the percent deviation between the values calculated in Procedure 11 and the resistances R_1 and R_2 measured in Procedure 2. Treat the values from Procedure 2 as the accepted values. (Recall that the percent deviation equals the positive difference between the experimental and accepted values divided by the accepted value time 100%.)

The percent deviation between R_1 and V_1/I_1 is _____ %.

The percent deviation between R_2 and V_2/I_2 is _____ %.

13. Calculate the equivalent resistance of the circuit by dividing the potential difference across the power supply (from Procedure 6) by the current through the power supply (from Procedure 8).

The equivalent resistance is $R_{eq} =$ _____ Ω.

14. Compute the reciprocals of the resistances R_1 and R_2 and the reciprocal of the equivalent resistance R_{eq} from Procedure 13.

$\dfrac{1}{R_1} =$ _____ Ω^{-1}

$\dfrac{1}{R_2} =$ _____ Ω^{-1}

$$\frac{1}{R_{eq}} = \underline{\hspace{5cm}} \Omega^{-1}$$

15. Sum the reciprocals of the two resistances and find the percent difference between this sum and the reciprocal of the equivalent resistance. Treat the sum as the "accepted value."

The percent deviation between $\frac{1}{R_1} + \frac{1}{R_2}$ and $\frac{1}{R_{eq}}$ is _____ %.

16. Sum the currents I_1 and I_2 through the two resistors (from Procedures 9 and 10) and find the percent deviation between this sum and the current through the power supply (from Procedure 8). Treat the power supply current as the "accepted value."

The percent deviation between $I_1 + I_2$ and the power supply current is _____ %.

GLOBAL QUESTIONS:

A. Was the current larger through the smaller or through the larger resistor?

Explain why this makes sense if the potential drop across each component is the same.

B. Lightbulbs are made of fine wires inside an evacuated glass bulb. When a lightbulb burns out, the wire breaks, leaving no path for the current. If you replace the two resistors you used with lightbulbs in this parallel circuit, would the good lightbulb remain lit if the other lightbulb burned out?

Why or why not?

C. In size, how does R_{eq} compare with R_1 and R_2? Was it larger than either of them, smaller than either of them, or somewhere between the two values?

D. Suppose you hooked a third resistor (R_3) in parallel with the two resistors you used in this experiment. In terms of R_1, R_2, and R_3 what would be the equivalent resistance of this new circuit? (*Hint:* Look at Procedures 14 and 15.)

E.   Would the current from the power supply be higher or lower when R_3 was added?

Why? (*Hint:* Use your answer from Question D and the definition of equivalent resistance to justify your answer.)

F.   Does adding resistors to a circuit change the potential supplied by the power supply? Does it change the current through any of the other resistors? Clearly explain your reasoning.

G.   (Optional) Are the outlets in your house connected in series or in parallel? How do you know?

H.   (Optional) Why are ammeters connected in series? Why do they have small resistances?

I.   (Optional) Why are voltmeters connected in parallel? Why do they have large resistances?

EXPERIMENT 28

THE GREENHOUSE EFFECT

Name_____   Partner(s)_____

Date_____   _____

Section Number_____   _____

BASE CONCEPTS: Temperature, Heat, Thermal Radiation

INTRODUCTION:

In large part because of the burning of fossil fuels such as oil and coal, the amount of carbon dioxide (CO_2) in the atmosphere is steadily increasing. The CO_2 traps thermal energy in the lower atmosphere, which *can* lead to global warming, a general increase in the earth's overall temperature, *if* there are no opposing atmospheric effects that promote global cooling. In turn, prolonged warming *could* lead to melting of portions of the two polar ice caps and major environmental changes. Such a warming phenomenon, caused by the increased CO_2, is commonly called the *greenhouse effect*. This laboratory will provide you an opportunity to observe the greenhouse effect in a simplified small-scale simulation.

SUPPLIES:

Two computerized temperature probes, dark soil, scale, metric ruler, small jar with a slit lid, incandescent lamp, small bowl.

In this lab you will investigate the temperature of the air above two soil samples subject to different conditions. One soil sample will be open to the room air, and the other will be enclosed in a glass jar.

PROCEDURE AND QUESTIONS:

1.   Using the scale, obtain two soil samples, each with a mass of 25 g.

2.   Place one soil sample in the small jar. Place the other sample in the small bowl.

3.   Place the lid on the jar and insert a temperature probe through the slit in the lid. Support the temperature probe so that it does not touch either the glass or the soil. Measure and record the distance between the temperature probe and the soil.

The end of the temperature probe is located _____ cm above the soil.

---

4.   Place the second temperature probe above the soil sample in the bowl, making sure its end is the same distance above the soil as the temperature probe's end in Procedure 3.

➤A.   Which of these configurations simulates the earth's atmosphere with increased CO_2?

       Why?

➤B.   What simulates the increased CO_2?

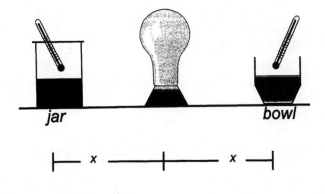

5.   Set the incandescent lamp about 30 cm away from each temperature probe. Make sure that the lightbulb is the same distance from each thermometer.

Figure 1

In the steps that follow you will be asked to record both temperatures for 8 minutes with the lamp on, and then for 8 more minutes with the lamp off. You should first *think* about what you expect to observe.

➢C.   After the first 8 minutes, just before you turn the lamp off, which of the two temperatures do you expect to be higher?

Why?

➢D.   After the light is off for 8 minutes, which of the two temperatures do you expect to be higher?

Why?

➢E.   How should the temperature of the jar and the bowl compare before the light is turned on? Why?

6.   Set your computer so that the temperature from each thermometer is recorded (every 30 seconds, if possible) for 16 minutes. After collecting data for 8 minutes, turn the light off. Use the data collected by the computer for the remainder of the lab.

7.   On the computer create a printed graph of temperature versus time for each of the temperature probes.

The Greenhouse Effect

GLOBAL QUESTIONS:

F.   Which system got hotter?

       Why?

G.   Did you predict this?

       Why or why not?

H.   Which system stayed hotter after the light was turned off?

       Why?

I.   Did you predict this?

       Why or why not?

J. Would you expect a cloudy night to be warmer or colder than a clear night if the days were the same temperature?

Why?

K. Describe a situation in which the greenhouse effect could be beneficial.

L. Describe a situation in which the greenhouse effect could be detrimental.

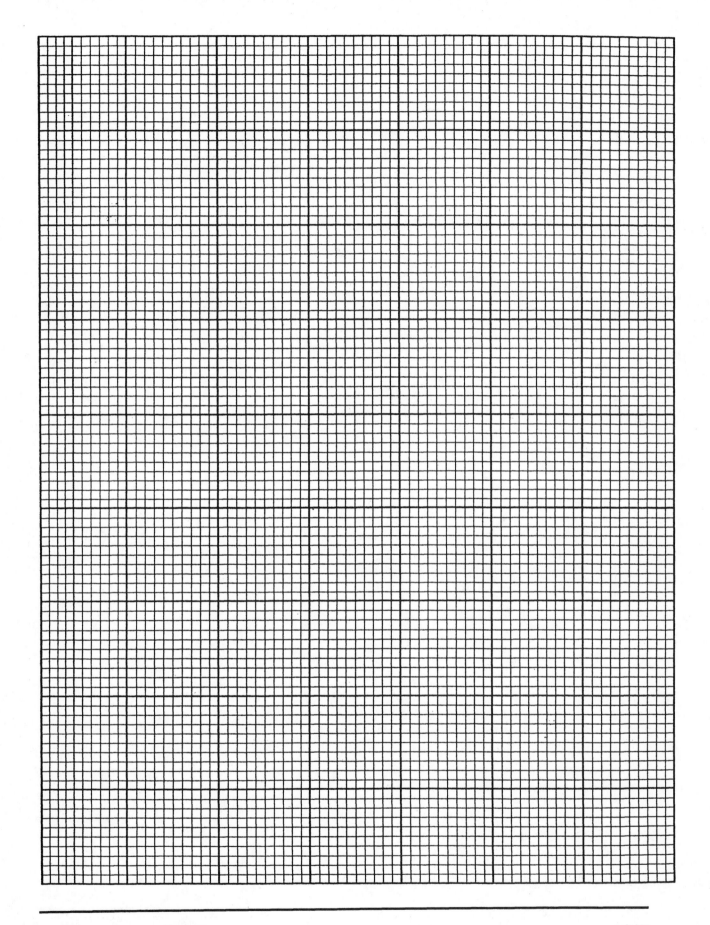

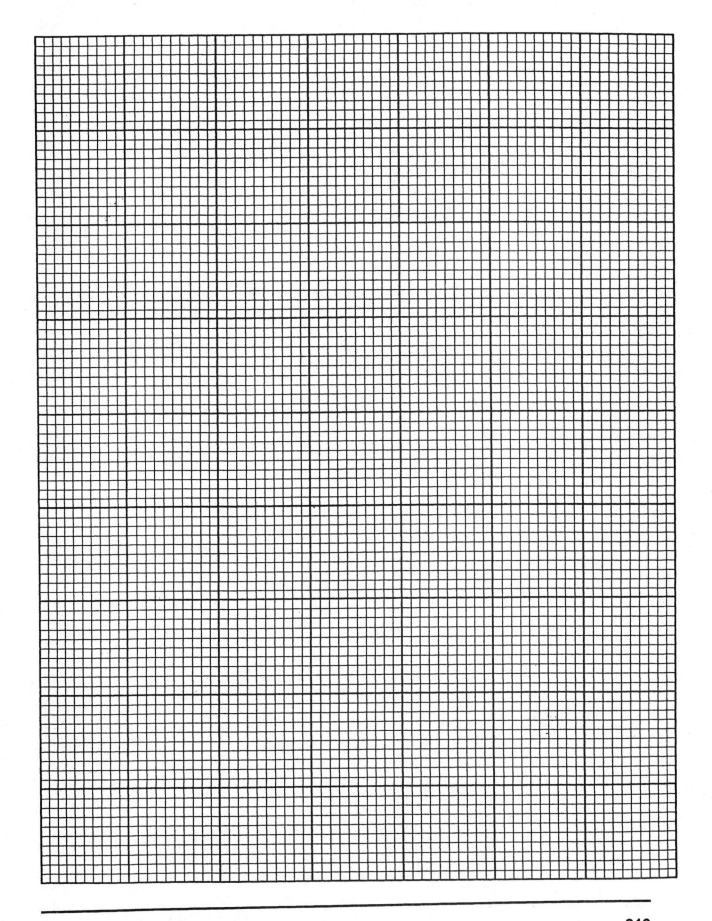

EXPERIMENT 29

ATOMS AND LIGHT: EMISSION SPECTRA

Name_____   Partner(s)_____

Date_____   _____

Section Number_____   _____

BASE CONCEPTS: Atomic Structure, Line Spectra

INTRODUCTION:

In this lab you will observe and analyze the visible radiation (light) from several sources by splitting the light from each source into its component wavelengths (colors). The set of wavelengths emitted by a light source is called the *spectrum* of the source. Each type of atom has a distinct spectrum, so spectra can be used as "fingerprints" for different elements. Spectroscopic experiments similar to this one have provided our basic knowledge of atoms and molecules. Likewise, much of our information about the heavens comes from spectral analysis of the electromagnetic radiation (including gamma rays, x-rays, and radio waves) coming to Earth from astronomical objects.

To investigate spectra from different sources, spectrometers are used. In general, spectrometers rely on interference of light waves to separate the wavelengths present. These distinct wavelengths appear as distinct lines called *spectral lines*. The qualitative nature of the spectrum can be discovered by looking at the brightness, color, and placement of the lines. In order to determine the actual wavelengths, the specific geometry of the spectrometer must be analyzed. There is a mathematical relationship between the wavelength, the spectrometer parameters, and the position of a spectral line. This relationship can be used to calibrate the spectrometer so that the position of a spectral line can be converted to a wavelength.

In this lab you will be looking at the emission of light from gaseous discharge tubes. The gas in a discharge tube is heated. This causes some of the atoms to jump to excited states. When an atom in an excited state relaxes back to a lower state, it emits its extra energy in the form of electromagnetic radiation. The amount of energy emitted dictates the wavelength of the radiation. The higher the energy, the shorter the wavelength. Because each type of atom has more than one excited state, radiation is emitted with several different wavelengths. This lab deals only with spectra in the visible light region.

SUPPLIES:

Diffraction grating spectroscope, diffraction grating, quantitative spectroscope, gaseous discharge tubes, incandescent lamp.

PROCEDURE AND QUESTIONS:

1.   Your instructor will tell you whether to work alone or in a group for this lab. Because there is no preferred sequence for this lab (Procedures 2–10 can be carried out in any order), you can easily share equipment.

➤A.   You are observing only the visible spectrum from these atoms. Do you suspect that there could be some lines in the nonvisible regions?

Why or why not?

        Note: Heated solids such as tungsten emit a continuous spectrum of light. In contrast, an isolated atom emits only those specific wavelengths in its fingerprint.

2.   **Incandescent bulb (tungsten) and large spectroscope:** When you look at this spectrum, notice that there are no distinct lines present. There is a continuous spectrum, as mentioned in the preceding paragraph. Notice the far ends of the spectrum. At one end, red shades into infrared, which is invisible to the naked eye. At the other end, violet shades into ultraviolet, which is also invisible to the naked eye. Measure and record the wavelength spread of the colors of the rainbow in units of nanometers (1 nm = 10^{-9} meter).

	violet	blue	green	yellow	orange	red
from						
to						

Remove the lid from the spectroscope to see how it works. Notice the positions of the slit and the grating. A sample grating is on the table nearby. Pick it up and look through it at the light until you can see rainbow colors in the grating.

---

3. **Discharge tubes and small spectroscopes:** You will see several sources, most of them labeled. Draw a picture of the observed spectra of mercury, helium, and hydrogen, roughly showing the color and position of each spectral line. Also, record the visual color of each source when looking at it without the spectroscope.

	violet blue green yellow orange red	Visual Color
Mercury Vapor		
Helium Gas		
Hydrogen Gas		

➤B. Hydrogen gas consists of two atoms bonded together, whereas mercury and helium gas are single atoms. Using this fact, explain the most obvious difference between the hydrogen spectrum and the spectra of the other two gases.

➤C. Describe in words the appearance of the spectra of the other labeled sources. Make sure you include air.

➤D.   Air consists mainly of two types of molecules. What are they?

      Both of the main types are two-atom molecules like hydrogen gas. This type of molecule
      is called *diatomic*. Explain why the spectrum of air is so complicated.

➤E.   By comparing spectra, decide whether the unlabeled source contains one of the labeled
      gases and, if so, which gas it contains. If you decide it does not contain any of the labeled
      gases, explain why you came to this conclusion.

      You have just done a spectroscopic chemical analysis!

4.   **Mercury vapor and large spectroscope:** Record the wavelengths (in nm) and colors of
      all the spectral lines you can observe in the mercury discharge tube, which is filled with
      vapor of the element mercury.

5.   **Helium gas with large spectroscope:** Record the wavelengths (in nm) and colors of all the spectral lines you can observe from the helium discharge tube.

6.   **Lightbulbs and small spectroscope:** Using a small spectroscope, compare the spectrum of an ordinary lightbulb with the spectrum of a fluorescent bulb.

➤F.   If you look closely, you should see some difference. What difference can you see? *Hint:* An ordinary lightbulb is a tungsten bulb like the one investigated in Procedure 2. A fluorescent bulb contains mercury vapor much like the discharge tube in Procedure 3; however, it also has a fluorescent coating that, after absorbing ultraviolet radiation emitted by the mercury atoms, re-emits light across the visible spectrum.

➤G.   Look at the spectra of the mercury discharge tube and then the fluorescent light. Do the two discrete spectra look alike? In what ways?

Atoms and Light: Emission Spectra

GLOBAL QUESTIONS:

H.   Suppose atoms of the hypothetical element Shazam have five possible "quantum states"; that is, five different energy levels. How many different energy "jumps" can you have from one state to another? Draw an energy level diagram showing these jumps. How many different wavelengths of light could you possibly see from this atom? Will the wavelengths emitted for the large energy jumps be shorter or longer than for the small energy jumps?

I.   For several years, two different types of vapor street lights have been widely used. These are merely gaseous discharge "tubes" built to emit a large amount of light.

What types of vapors (i.e., what chemical elements) are used in these lights?

If you don't recall the appearance of these lamps, look for each type some evening. Describe the visual appearance of the light coming from each.

Judging from the appearance of their light, what differences would you expect to see in the spectra of these two lamps?

J.   Explain how astronomers have been able to study the chemical composition of the planet Jupiter even though no manned space flight has ever gone to Jupiter.

EXPERIMENT
30

THE HUBBLE CONSTANT

Name_____   Partner(s)_____

Date_____   _____

Section Number_____   _____

BASE CONCEPTS: Graphing

INTRODUCTION:

During much of this century there has been heated debate as to whether the universe is more or less static in nature or expanding. Today nearly all scientists accept the big bang theory of an expanding universe. This theory holds that the galaxies that make up the universe are all rapidly moving away from one another. The speeds at which other galaxies are moving away from our galaxy, the Milky Way, can be determined by examining their light spectra. There are matches between the spectra of known elements and the galactic spectra, but all the spectral lines from distant galaxies are shifted to longer wavelengths. This effect is known as the cosmological *red shift*.

Because there is an inverse relationship between the wavelength of light and its frequency, the light is shifted to lower frequencies. This frequency decrease is due to the *Doppler effect*, the phenomenon that, for sound, causes you to hear a siren at a lower frequency when you and the siren are moving away from each other. By using the known spectra for hydrogen and other elements, astronomers can relate the size of the shift to the speed at which the galaxy is moving away from us.

Astronomers can also estimate the distance from the Milky Way to the galaxies, though with much lower accuracy, by using techniques to estimate the *intrinsic* brightness of stars and galaxies and then using the *observed* brightness to deduce the distance.

In 1929 Edwin P. Hubble published his observation that the recession speeds of observed galaxies are proportional to their distances from the Milky Way (*Hubble's law*). In mathematical terms, $V = H \times D$, where V is the galaxy's recession speed, D is the distance from the Milky Way to the galaxy, and H is the Hubble constant. If V is measured in km/s and D in Megaparsecs (Mps), H has been determined to be between 50 and 100 (km/s)/Mps, depending on the specific observational data used. In this lab you will investigate a simple one-dimensional elastic analog that illustrates the way this expansion leads to Hubble's law.

SUPPLIES: Elastic strip about 50 cm long, safety pins, tape measure.

PROCEDURE:

1. Pin six safety pins along the elastic strip at more or less random positions, leaving at least 2 in. between adjacent pins.

2. Stretch the elastic strip to a length about one and a half times its original length. Figure out a way to keep it this length while you are measuring distances.

3. Designate one of the six pins as *the Milky Way* and measure the distances from *the Milky Way* pin to each of the other five pins. Take all distances to be positive, regardless of direction. Make sure that you record distances in the **D** column in Table 1.

4. Stretch the elastic strip to a length about twice its original length. Again measure the distances from the Milky Way pin (the same pin you chose in Procedure 3 for the Milky Way) to the other five pins, again taking all distances as positive. Record these distances in the **D′** column of Table 1.

Table 1
Data from First Trial
(Draw a dash rather than entering a number for the Milky Way pin.)

Pin #	D (cm) Initial Distance	D′(cm) Final Distance (cm)	V (cm/s) Recession Velocity
1			
2			
3			
4			
5			
6			

5. Enter the values in the D' and D columns into a spreadsheet.

6. Calculate the "recession velocity" V by assuming that your "universe" has stretched from the size in Procedure 3 to the size in Procedure 4 in 1 second. In this time, each "galaxy" pin has moved with respect to the Milky Way from distance D to distance D'; so V equals the difference $(D' - D)$ divided by 1 second (the time it took to move that distance). Mathematically,

$$V = (D' - D)/(1 \text{ s}) = (D' - D)/\text{s}$$

7. Plot a graph of V versus D.

8.   Draw a recession line (the best straight line that comes the closest to the most data points) through the data points on the graph.

9.   Determine your elastic "Hubble constant" by finding the slope of the regression line from Procedure 8.

10.   Repeat Procedures 2–9, choosing a different pin as the Milky Way. Enter this information in Table 2

Table 2
Data from Second Trial
(Draw a dash rather than entering a number for the Milky Way pin)

Pin #	D (cm) Initial Distance	D' (cm) Final Distance (cm)	V (cm/s) Recession Velocity
1			
2			
3			
4			
5			
6			

GLOBAL QUESTIONS:

A.   Compare your two graphs. Do they look similar? Is the slope basically the same for both? Is the vertical axis intercept (the value of V when D is zero) the same on both? If so what is the value? If not, what are the two values?

B.   Did the more distant galaxies move a greater distance in your universe? Support your answer with data from the table and your graph.

C.   Suppose Galaxy HX-43 is located 100 cm from the Milky Way in your universe. What is its recessional velocity?

D.   What would a negative Hubble constant mean?

E.   From your experience with sound, how does the frequency of a siren change when you are moving *toward* the siren? If the galaxies moved toward us, how would the appearance of the spectral lines change?

F.   Given the following relationships (conversion equations): 1 km = 1000 m, and 1 megaparsec (Mps) = 3.09 X 10^{22} m, rewrite Hubble's constant $H = 100$ km/s/Mps in units of 1/s.

G.   Explain what the "Hubble's constant" for your elastic strip means.

H.   Suppose you had an elastic strip that was much easier to stretch than the one you used. It was the exact length and stretched the same amount as in this experiment. The pins were put in the same spots and the same pins were chosen from the Milky Way. What differences, if any, would you expect in your collected data?

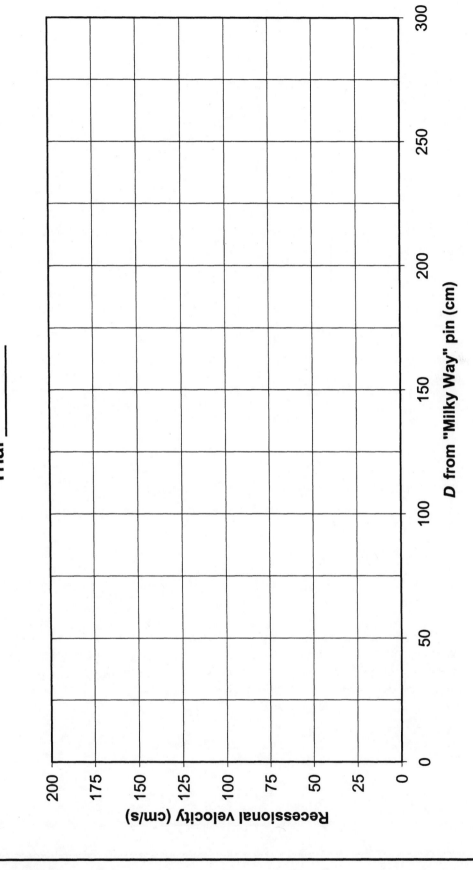

Hubble's Constant Activity

Trial _____

Recessional velocity (cm/s)

D from "Milky Way" pin (cm)

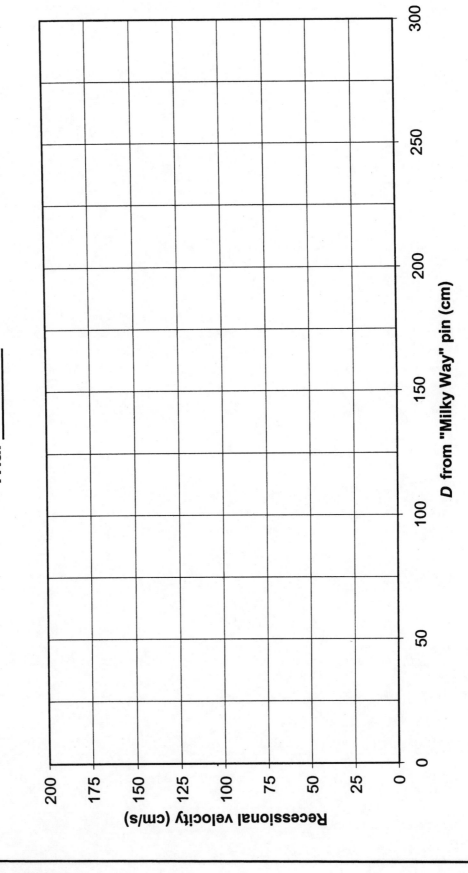

Hubble's Constant Activity
Trial ____

Recessional velocity (cm/s)

D from "Milky Way" pin (cm)

EXPERIMENT
31

RADIOACTIVITY AND PENNIES

Name_____   Partner(s)_____

Date_____   _____

Section Number_____   _____

BASE CONCEPTS: Nuclear Atom

INTRODUCTION:

The atomic nuclei (plural of *nucleus*) of radioactive substances go from an unstable state to a stable state by emitting energetic particles. This process is called *radioactive decay*. For a given type of nucleus, in each unit of time (e.g., a second) there is a certain statistical probability that *each* nucleus will decay. The number of decays per unit of time in a radioactive sample (usually called the *activity* of the sample) is therefore proportional to this probability and to the number of unstable nuclei remaining in the sample. A process of this kind is characterized by the fact that the average time it takes for half of the nuclei to decay is *independent* of the original number of nuclei. This time is called the *half-life* of the radioactive substance (or, equivalently, of that type of nucleus). The *half-life* is often used to identify radioactive substances. It also plays a crucial role in determining the ages of many archaeological and geological samples.

In this lab exercise you will simulate a radioactive decay process by tossing a large sample of pennies and removing those that come up tails on each toss.

SUPPLIES: Coins, container.

PROCEDURE AND QUESTIONS:

Before starting to collect data for this activity, you will find it helpful to study some idealized data of this type. Assume that you start with a radioactive sample of 20,000 unstable nuclei with a half-life of 2 seconds. Unlike real nuclei, these hypothetical nuclei decay in an ideal fashion with no statistical fluctuations. Table 1 shows how the number of these nuclei in the sample changes with time.

1.   In a spreadsheet or a graphic calculator, enter the data from Table 1 and create a graph.

Table 1
Example Data

TIME (s)	NUMBER OF NUCLEI REMAINING
0	20,000
2	10,000
4	5,000
6	2,500
8	1,250
10	625
12	312
14	156
16	78
18	39

2. You can determine the half-life simply by finding the time it takes for half of the nuclei to decay. Because of the nature of the decay process, you may choose any starting number of radioactive nuclei, divide that number by 2, find the times at which these two numbers of nuclei were present, and subtract these times. Try this with the graph you made for the data in Table 1, by extrapolating from the graph.

The time corresponding to 16,000 nuclei present is _____ seconds.

The time corresponding to 8000 nuclei present is _____ seconds.

The time corresponding to 4000 nuclei present is _____ seconds.

➢A. From the preceding information, the half-life of this type of nucleus is _____ seconds.

How do you know?

Now you are ready to collect and examine some data that demonstrate the definition of half-life.

3. Start with a sample of 300 pennies (the initial radioactive "nuclei" in this simulation). Record this initial number in Table 2. Place these coins in a container.

4. Shake the container a few times, and "toss" the coins out on the table top. Each toss represent a certain amount of time. Put the coins that came up **tails** off to the side. These represent the nuclei that decayed in the time represented by the toss. Count the number of coins that came up **heads**.

5. In Table 2, record the number of **heads** (remaining coins, or "remaining nuclei") after the first throw. Place only the coins that came up **heads** back in the container.

Table 2
Penny Data

NUMBER OF THROWS	NUMBER OF COINS REMAINING
0	Initial # =
1	
2	
3	
4	
5	
6	
7	
8	
9	
10	
11	
12	

6. Repeat Procedures 4 and 5 with the remaining coins until there are fewer than four coins left. (There are probably more rows in Table 2 than you need. Just leave any extra rows blank.)

➤B.   In time units of "number of throws," what do you expect the half-life of the pennies to be?

Why?

7.   Make a graph (in a spreadsheet or by hand) of the number of pennies remaining versus number of throws ("time").

8.   Draw a smooth curve that passes through or near the data points on the graph. Following the process described earlier, find the half-life (in "throws").

The half-life of the pennies is _____ throws.

➤C.   How does this value compare with your prediction in Question B?

GLOBAL QUESTIONS:

D.   Suppose that rather than tossing pennies you toss dice to collect data. You retain all dice that come up 1 and put aside all the dice which come up 2, 3, 4, 5, or 6. Will it take fewer throws or more throws (compared with the same initial number of pennies) for all the dice to be put aside?

How do you know?

E.   For the dice, will the half-life (in "throws") be larger or smaller than the half-life for the pennies?

How do you know?

F.   How does the half-life depend on the number of pennies (or atoms, or dice, etc.) with which you start?

G.   How does your ability to calculate the half-life depend on the number of pennies (or atoms, or dice, etc.) you have to begin with? Give an example.

H. What is the half-life of a radioactive substance if 1/16 of the original amount is left after 20,000 years?

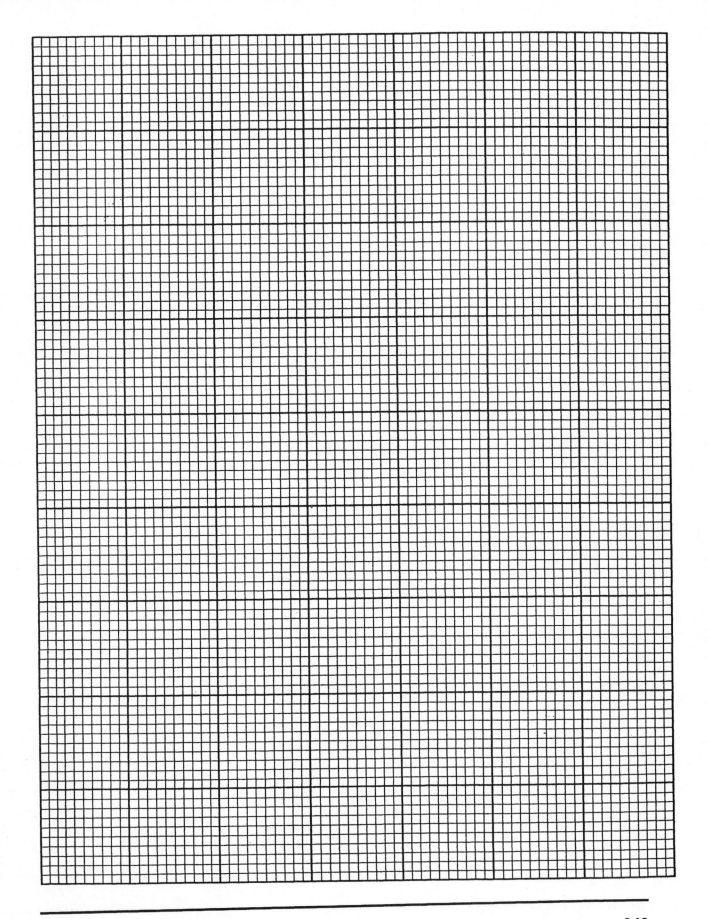

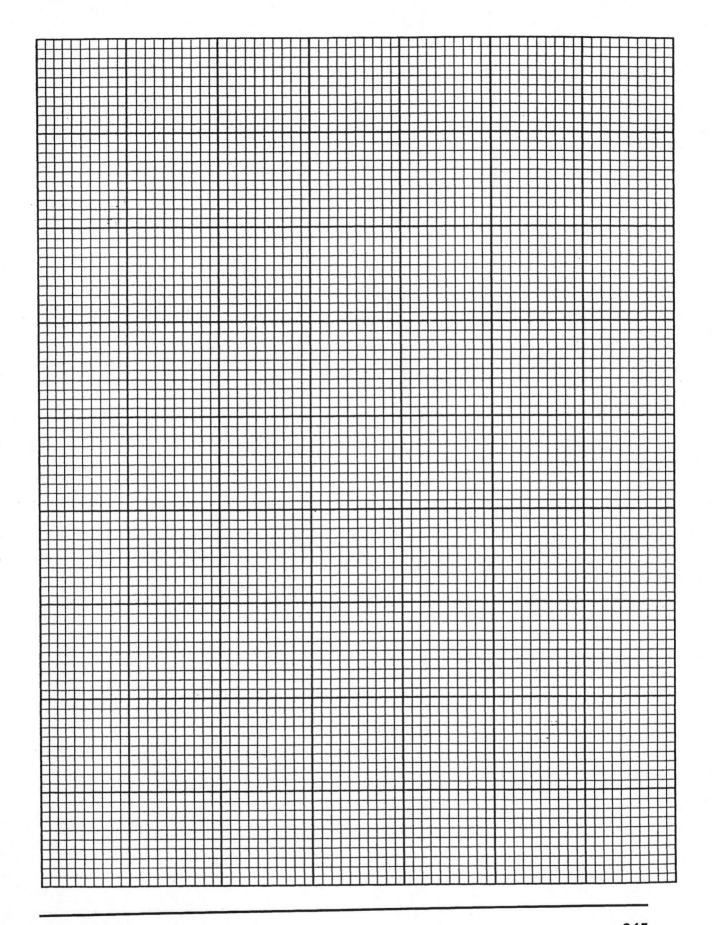

ADDITIONAL GRAPH PAPER

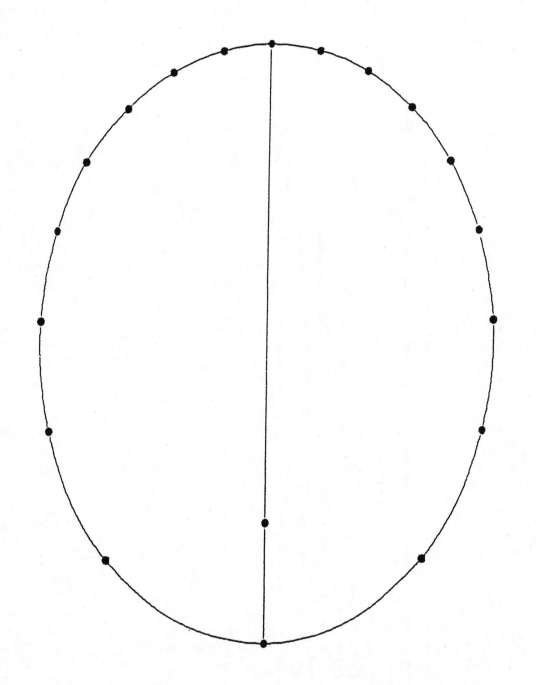

Temperature as a Function of Time

Temperature (degrees Celsius)

90
80
70
60
50
40
30
20

0   5   10   15   20

Time (minutes)

Hubble's Constant Activity

Trial ________

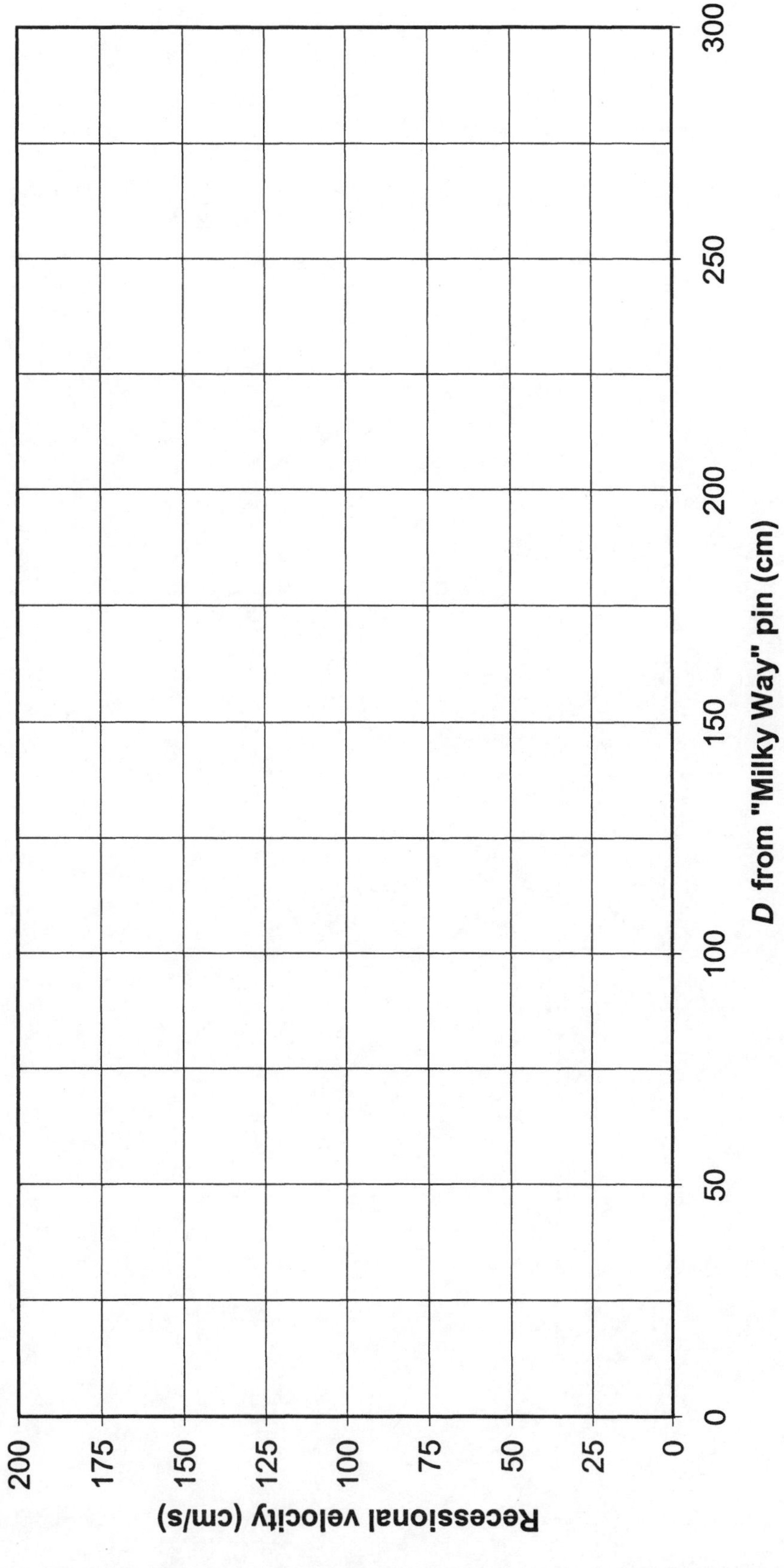

200
175
150
125
100
75
50
25
0
Recessional velocity (cm/s)
0
50
100
150
200
250
300
D from "Milky Way" pin (cm)

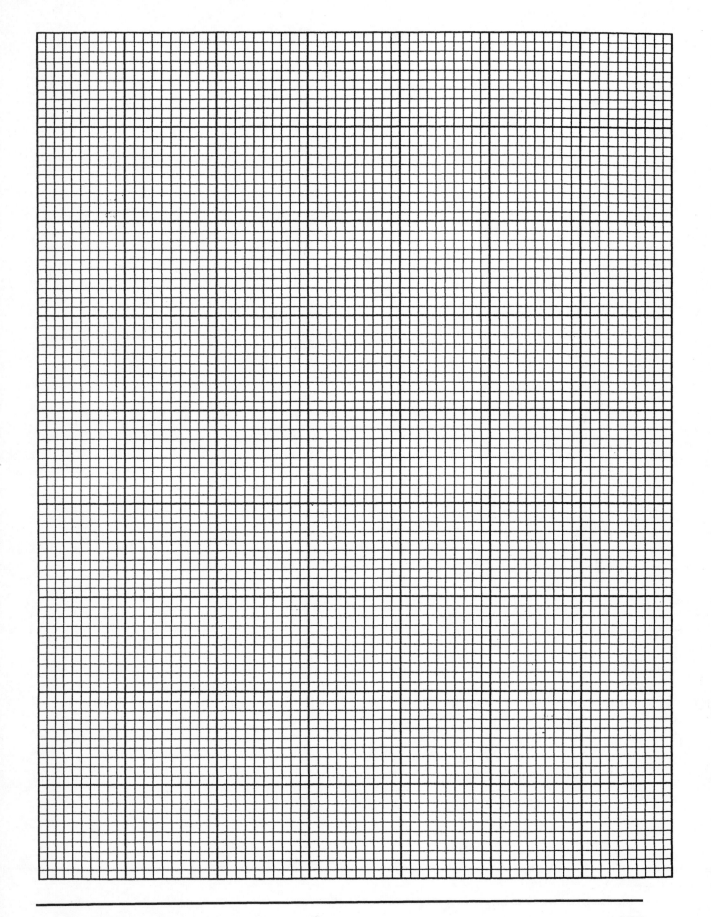

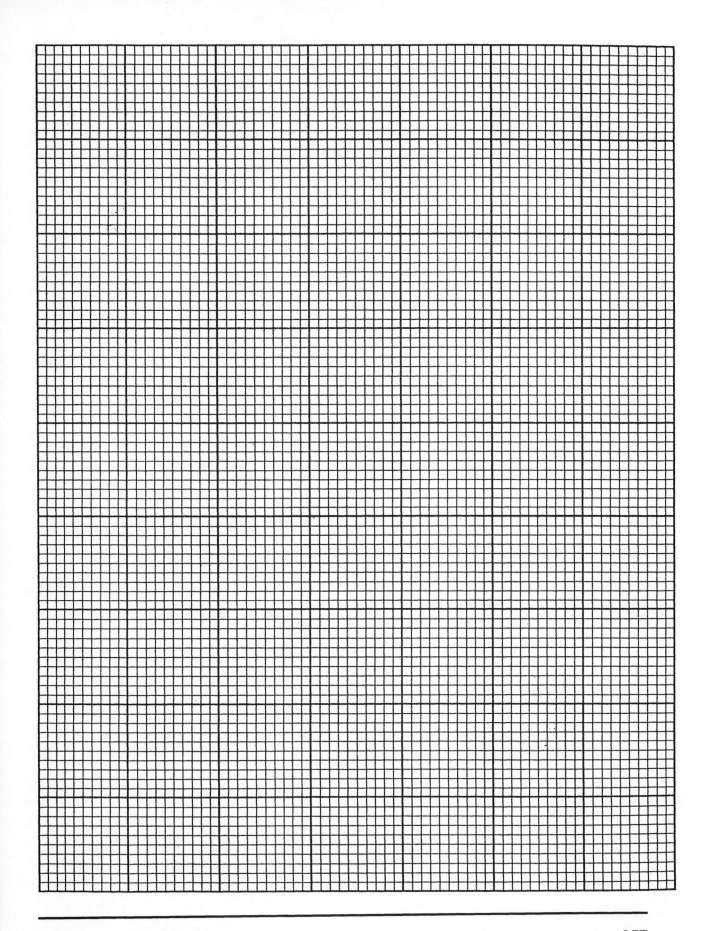

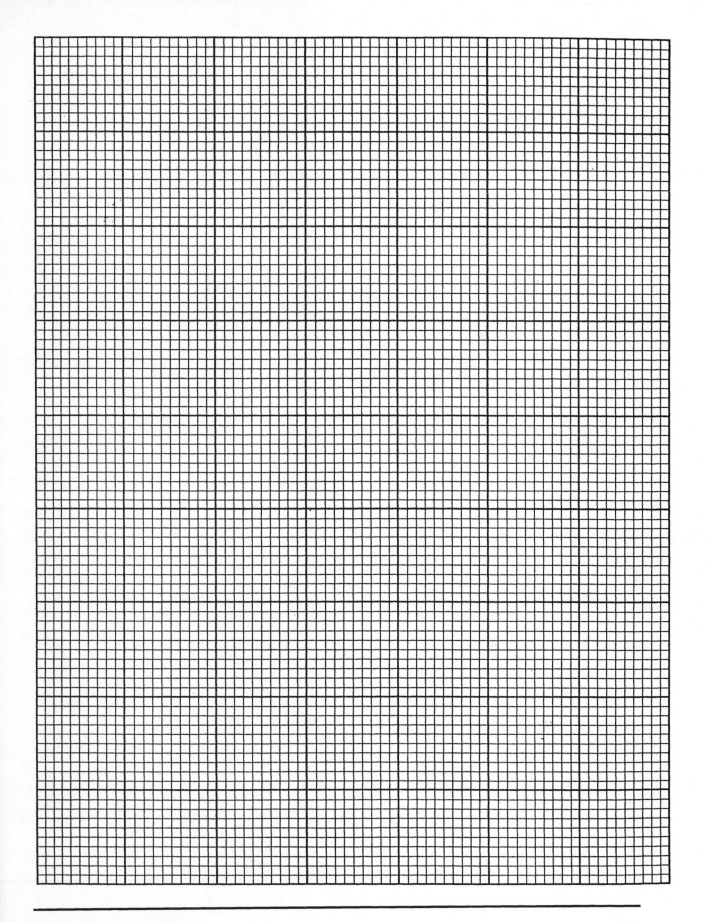